A BASIC MATH APPROACH TO CONCEPTS OF CHEMISTRY

Sixth Edition

A BASIC MATH APPROACH TO CONCEPTS OF CHEMISTRY

Sixth Edition

LEO MICHELS

Milwaukee Area Technical College

Brooks/Cole Publishing Company

I(T)P™ An International Thomson Publishing Company

Pacific Grove • Albany • Bonn • Cincinnati • Detroit • London • Madrid • Melbourne • Mexico City
Mexico City • New York • Paris • San Francisco • Singapore • Tokyo• Toronto • Washington

Editor: *Faith B. Stoddard*
Editorial Assistant: *Beth Wilbur*
Production Editor: *Penelope Sky*
Advertising: *Romy Fineroff*
Marketing: *Connie Jirovsky*

Cover Design: *Laurie Albrecht*
Cover Photo: *Carr Clifton*
Typesetting: *CompuKing Typesetting*
Printing and Binding: *Patterson Printing*

For more information, contact:

BROOKS/COLE PUBLISHING COMPANY
511 Forest Lodge Road
Pacific Grove, CA 93950
USA

International Thomson Editores
Campos Eliseos 385, Piso 7
Col. Polanco
11560 México D. F. México

International Thomson Publishing Europe
Berkshire House 168-173
High Holborn
London WC1V 7AA
England

International Thomson Publishing GmbH
Königswinterer Strasse 418
53227 Bonn
Germany

Thomas Nelson Australia
102 Dodds Street
South Melbourne, 3205
Victoria, Australia

International Thomson Publishing Asia
221 Henderson Road
#05-10 Henderson Building
Singapore 0315

Nelson Canada
1120 Birchmount Road
Scarborough, Ontario
Canada M1K 5G4

International Thomson Publishing Japan
Hirakawacho Kyowa Building, 3F
2-2-1 Hirakawacho
Chiyoda-ku, Tokyo 102
Japan

Printed in the United States of America
10 9 8 7 6 5 4 3 2

PREFACE

To the Instructor

The purpose of this textbook is to teach the basic concepts of chemistry and problem solving techniques to students who may not have science backgrounds or extensive math skills. The book can be used as a main text for a preparatory course or as a supplement in survey and general chemistry courses. Many chemistry students who need help outside the classroom have found it useful as a self-paced learning manual.

I have greatly expanded the number of topics since the first edition, including more complex problems that involve weight relations, stoichiometry, solutions, and gas laws; I have also illustrated the use of the calculator. Dimensional analysis and conversion factors in problem solving are presented in a manner that facilitates students' understanding of each step.

In this edition I have introduced SI units and used current abbreviations for measurement units throughout. The conversion factor method of problem solving is emphasized, and there are more problems in the unit on balancing equations.

The topics in each unit progress from basic to complex, and those at the end of each unit can be omitted. The first five units are devoted to number concepts, measurement concepts, and the measurement systems that are used in science. The remaining nine units are concerned with the division and properties of matter, atomic structure, chemical bonding and formula writing, nomenclature, equation balancing, weight relations, stoichiometry, solutions, and gas laws.

The physical format of the book invites active student involvement in the learning process. Dividing the material into frames and using questions within them encourages the student to master each new idea before moving on to the next one; the answers are on the same page as the questions, so students know immediately whether they are ready to progress. At the end of each unit are self-evaluation tests, with the parts of the unit covered by each test clearly marked.

To the Student

The purpose of this text is to allow you to master basic concepts and skills. The first six units present the math skills, problem solving techniques, and measurement systems you'll need to know in order to do the work in the remaining nine units. Each unit is divided into parts; each part is concerned with a major concept. The parts consist of a series of frames that contain the individual ideas on which the major concept is based. Each idea is first explained and then illustrated with an example. You are then asked to use the idea to solve problems. The answers appear on the same page as the problems, so you will know immediately whether your response is correct. The answers are staggered: the answer to frame 1 is beside frame 2, the answer to frame 2 is beside frame 3, and so on. Where an answer is too long to fit in the answer column, a separate answer frame has been inserted after the frame that contains the question.

Self-evaluation tests are included at the end of each unit. The section of the unit covered by each test is clearly marked. You can use the tests to check how well you have learned the material. The answers to the tests are at the back of the book.

As you work the problems and answer the questions in the frames, remember that incorrect answers usually indicate that you have not fully grasped the material. You should reread the frame and try again. When you get to the self-evaluation tests you should have little difficulty with them.

Acknowledgments

I am grateful for the help given over the years by colleagues who have used this book and suggested many improvements.

I appreciate the support of the staff at Brooks/Cole Publishing Company. In particular, I thank Penelope Sky and Faith Stoddard.

I am indebted to Dr. Morris Hein, who proofread the original manuscript, and to the following reviewers, who offered many helpful comments: Wade A. Freeman, University of Illinois–Chicago; Catharine Hickey-Williams, Western Connecticut State University; and John G. Russell, California State University–Sacramento.

Leo Michels

CONTENTS

A BASIC MATH APPROACH TO CONCEPTS OF CHEMISTRY

Sixth Edition

UNIT 1
Whole Numbers and Decimals

In this unit the naming of whole numbers and decimals will be reviewed. You will learn the place name digits of whole numbers and decimals, how to change a number from its word name to its numerical form and how to add, subtract, multiply, and divide decimal numbers.

Part 1. Place Names of Whole Numbers

1. There are ten digits that make up our number system: 0, 1, 2, 3, 4, 5, 6, 7, 8, and 9. All numbers are made up of these digits.

   Example: 397 is a three-digit number

   41,646 is a _____ -digit number

2. A three-digit number has three places. Names are given to these places as follows:

   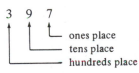

   In the number 428, state what digit is in the

   a. ones place _____

   b. tens place _____

   c. hundreds place _____

1. five

3. In a six-digit number the places are named as follows:

```
        4  8  2 , 3  5  8
hundred-thousands ─┘  ↑  ↑   ↑ ↑ └─ ones
ten-thousands ───────┘  │   │ └─── tens
thousands ──────────────┘   └───── hundreds
```

In the number 567,123 state what digit is in the indicated place.

a.   thousands place _____

b.   ten-thousands place _____

c.   hundred-thousands place _____

2. a. 8
   b. 2
   c. 4

4. A nine-digit number is named as follows:

```
        9  8  7 , 5  4  6 , 3  2  1
hundred-millions ─┘  ↑  ↑   ↑     ↑ ↑ └─ ones
ten-millions ───────┘  │   │     │ └─── tens
millions ──────────────┘   │     └───── hundreds
hundred-thousands ─────────┘          └─ thousands
                    ten-thousands
```

In the number 123,456,879, state what digit is in the indicated place.

a.   hundred-millions place _____

b.   ten-millions place _____

c.   millions place _____

3. a. 7
   b. 6
   c. 5

5. In the number 407,639,582, state what digit is in the indicated place.

a.   ten-millions place _____

b.   ten-thousands place _____

c.   tens place _____

d.   hundred-thousands place _____

4. a. 1
   b. 2
   c. 3

Part 2. Word Names of Large Whole Numbers

1. When naming three-digit numbers, the number of hundreds is named first, followed by the name of a number from 1 to 99.

   Example: 342 is named three hundred, forty-two

                   number of   number from
                   hundreds   1 to 99

   Name these numbers.

   a.   947 _____

   b.   289 _____

5. a. 0
   b. 3
   c. 8
   d. 6

2. Write the number that goes with the word name.

    a.   four hundred sixty _____

    b.   nine hundred nineteen _____

    c.   six hundred _____

    d.   three hundred four _____

3. When naming numbers in the thousands (thousands, ten-thousands, and hundred-thousands), we name only how many thousands are present; we never name the number of ten-thousands or hundred-thousands separately.

    Example: 847,000 is called eight hundred forty-seven thousand

    Name these numbers.

    a.   567,000 _____

    b.   234,000 _____

    c.   82,000 _____

    d.   7000 _____

4. Write the number that goes with each word name.

    a.   eighteen thousand _____

    b.   six hundred seven thousand _____

    c.   eight thousand _____

    d.   nine hundred thousand _____

5. We can name six-digit numbers by using the rules from frames 1 and 3 of Part II.

    Example: 587,486 is called five hundred eighty-seven thousand, four hundred eighty-six

    Name these numbers.

    a.   623,145

    _____

    b.   897,085

    _____

6. Write the number that is indicated by the word name.

    a.   four hundred sixteen thousand, twenty-eight _____

    b.   eight hundred thousand, two hundred forty-nine _____

    c.   nine hundred thirty-three thousand, eighteen _____

    d.   sixty-two thousand, one hundred eleven _____

---

1. a. nine hundred forty-seven
   b. two hundred eighty-nine

2. a. 460
   b. 919
   c. 600
   d. 304

3. a. five hundred sixty-seven thousand
   b. two hundred thirty-four thousand
   c. eighty-two thousand
   d. seven thousand

4. a. 18,000
   b. 607,000
   c. 8000
   d. 900,000

5. a. six hundred twenty-three thousand, one hundred forty-five
   b. eight hundred ninety-seven thousand, eighty-five

7. When naming numbers in the millions (millions, ten-millions, hundred-millions), we name how many millions are present; we never name the number of ten-millions or hundred-millions separately.

    Example: 234,000,000 is called two hundred thirty-four million

    Name these numbers.

    a.   168,000,000 _____

    b.   92,000,000 _____

    c.   7,000,000 _____

8. Using the rules we have learned previously, we can name nine-digit numbers.

    Example: 543,286,492 is named five hundred forty-three million, two hundred eighty-six thousand, four hundred ninety-two

    Name these numbers.

    a.   827,462,123

    _____

    b.   901,000,842

    _____

    c.   8,100,009

    _____

9. Write the number that is indicated by the word name.

    a.   three hundred eighty-two million, four hundred thirty-two thousand, nine hundred fifty-six _____

    b.   forty-three million, three hundred thousand, seventy-two _____

    c.   two million, nine thousand, ten _____

10. In the number 905,643,182, state what digit is in the indicated place.

    a.   thousands place _____   b.   tens place _____

    c.   hundred-millions place _____   d.   ones place _____

    e.   ten-millions place _____   f.   hundreds place _____

    g.   hundred-thousands place _____   h.   millions place _____

6. a. 416,028
   b. 800,249
   c. 933,018
   d. 62,111

7. a. one hundred sixty-eight million
   b. ninety-two million
   c. seven million

8. a. eight hundred twenty-seven million, four hundred sixty-two thousand, one hundred twenty-three
   b. nine hundred one million, eight hundred forty-two
   c. eight million, one hundred thousand, nine

9. a. 382,432,956
   b. 43,300,072
   c. 2,009,010

11. Write the number indicated by the word name.

    a. eight hundred sixty-six _____

    b. four hundred four _____

    c. nine thousand, two hundred _____

    d. twenty-eight thousand, forty _____

    e. one hundred eighteen thousand, four hundred twelve _____

    f. three million, eight hundred thousand _____

    g. twenty-two million, nine hundred forty-two _____

    h. seven hundred fifty-six million, three hundred nineteen thousand, seven hundred

       four _____

10. a. 3   b. 8
     c. 9   d. 2
     e. 0   f. 1
     g. 6   h. 5

12. Answers to frame 11.

    a. 866   b. 404   c. 9200   d. 28,040
    e. 118,412   f. 3,800,000   g. 22,000,942   h. 756,319,704

Part 3. Place Names of Decimal Numbers

1. A decimal number is a number whose value falls between two whole numbers. For example, the number 4.5 is a decimal number. It represents a value greater than 4 and less than 5. A decimal number with a value between 0 and 1 is represented by a 0 to the left of the decimal point and the decimal value to the right of the decimal point.

    Examples: 0.1, 0.06, 0.0102

    Which of the following are decimal numbers? _____

    a. 123   b. 0.006   c. 2.0106   d. 1089

2. Just as in whole numbers, the digits in decimal numbers have place names.

    Example: 0 . 0 1 3

              ┌─ thousandths
            ├─ hundredths
          └─ tenths

    Note that the place-name endings of digits to the right of the decimal point end in *ths*. The place-name endings of digits in a whole number end in the letter *s*. Also note that in a decimal number there is no place to the right of the decimal point corresponding to the ones place, as there is in a whole number; the first place to the right of the decimal place is the tenths place.

    In the number 0.014, state what digit is in the indicated place.

    a. tenths place _____

    b. hundredths place _____

    c. thousandths place _____

1. b and c

3. Place names to the right of the decimal point end with the letters _____.

2. a. 0   b. 1   c. 4

4. The first place to the right of the decimal point in a decimal number is called the _____ place.

3. ths

5. The six digits to the right of the decimal point are named as follows.

In the decimal number 0.000876, state what digit is in the indicated place.

a. ten-thousandths place _____

b. hundred-thousandths place _____

c. millionths place _____

4. tenths

6. In the decimal number 0.100206, state what digit is in the indicated place.

a. tenths place _____

b. hundredths place _____

c. ten-thousandths place _____

d. millionths place _____

5. a. 8
   b. 7
   c. 6

Part 4. Naming Decimal Numbers

1. The rule for naming decimal amounts is: give the word name of the number and add the place name of the last digit.

6. a. 1
   b. 0
   c. 2
   d. 6

Examples: 0.27   = twenty seven hundredths
                       name of   place name
                       number   of last digit

0.462   = four hundred sixty-two thousandths
0.16   = sixteen hundredths
0.0082 = eighty-two ten-thousandths

Name these numbers.

a. 0.9 _____

b. 0.35 _____

c. 0.002 _____

d. 0.0096 _____

2. When writing the decimal number from the word name, we apply the rule in reverse. Put the last digit of the number in the place named.

Example: twenty-two thousandths = 0.022

         last digit   place where
         of number   last digit goes

seven ten-thousandths = 0.0007
eight hundred nine thousandths = 0.809
sixteen hundredths = 0.16

Write the decimal number indicated by the place name.

a.   four tenths _____

b.   eight hundredths _____

c.   thirty-two thousandths _____

d.   nineteen ten-thousandths _____

3. Decimal numbers larger than one can be named by going through the following steps.

a.   Name the whole number.
b.   Use the word *point* for the decimal point.
c.   Give the names of the individual digits of the decimal amount from left to right.

Example: 22.37   =   twenty-two   point   three seven

         name of   decimal   names of
         whole number   point   decimal
                             digits

    8.09   = eight point zero nine
    80.0   = eighty point zero
  50.004 = fifty point zero zero four

Name these numbers.

a.   6.2 _____

b.   9.07 _____

c.   51.06 _____

d.   88.0016 _____

4. Write the decimal number indicated by the name.

a.   seven point zero _____

b.   eighteen point one two _____

c.   nine point zero two _____

d.   twenty-seven point zero zero six _____

---

1. a.   nine tenths
   b.   thirty-five hundredths
   c.   two thousandths
   d.   ninety-six ten-thousandths

2. a.   0.4
   b.   0.08
   c.   0.032
   d.   0.0019

3. a.   six point two
   b.   nine point zero seven
   c.   fifty-one point zero six
   d.   eighty-eight point zero zero one six

5. Solve the problems below.

In the number 0.00287, state what digit is in the indicated place.

a.   hundredths place _____

b.   ten-thousandths place _____

Write the name of the decimal numbers.

c.   0.8 _____

d.   0.16 _____

e.   0.008 _____

f.   0.562 _____

Write the decimal number indicated by the name.

g.   six tenths _____

h.   three hundredths _____

i.   sixty-six thousandths _____

j.   eight ten-thousandths _____

Name the decimal numbers.

k.   9.3 _____

l.   7.07 _____

m.   72.01 _____

n.   6.0011 _____

Write the decimal number indicated by the name.

o.   one point two _____

p.   twenty-one point zero _____

q.   eighty-nine point zero seven _____

r.   eleven point zero zero one six _____

Part 5. Adding and Subtracting Decimal Numbers

1. Decimal numbers are added just as whole numbers are. Care must be taken that all of the digits in any one number place are lined up properly.

Example: tenths column
                     hundredths column

        0.23
      + 0.44

Note that all of the digits in the tenths place are lined up in a column and digits in the hundredths place are also in a column. When the decimal points are lined up one beneath the other, the number places are properly aligned.

Examples: $0.035 + 0.009 \rightarrow$ 0.035 $11.009 + 19.123 \rightarrow$ 11.009
                              + 0.009   + 19.123

Write the decimal numbers in proper form for addition. (Do not solve.)

a.   0.103 + 0.006   b.   13.123 + 10.001

---

4. a. 7.0
   b. 18.12
   c. 9.02
   d. 27.006

5. a. 0
   b. 8
   c. eight tenths
   d. sixteen hundredths
   e. eight thousandths
   f. five hundred sixty-two
        thousandths
   g. 0.6
   h. 0.03
   i. 0.066
   j. 0.0008
   k. nine point three
   l. seven point zero seven
   m. seventy-two point zero one
   n. six point zero zero
        one one
   o. 1.2
   p. 21.0
   q. 89.07
   r. 11.0016

2. Write these decimal numbers in proper form for addition and find the correct answer.

   a.   0.807 + 0.116

   b.   1.76 + 4.13

3. There are times when the decimal numbers being added have different numbers of digits.

   Example: 1.6
   $$\quad\quad + \underline{2.03}$$

   The difference in the number of digits can often cause confusion. Care must be taken to line up the decimal digits in their proper places by aligning the decimal points and the numbers on each side of the decimal points.

   Examples: 11.6 + 2.03 →   11.6
   $$\quad\quad\quad\quad\quad + \underline{\quad 2.03}$$

   6.025 + 22.1 + 1.0016 →   6.025
   $$\quad\quad\quad\quad\quad\quad\quad\quad\quad 22.1$$
   $$\quad\quad\quad\quad\quad\quad + \underline{\quad 1.0016}$$

   Rewrite these addition problems so that the digits line up properly. (Do not solve.)

   a.   0.02 + 0.1006 + 0.9

   b.   1.8 + 10.09 + 3.001

4. Rewrite these problems and solve them.

   a.   1.01 + 10.1 + 12.006

   b.   22.007 + 9.12 + 31.8

5. The same precautions must be taken when subtracting decimal numbers. All digits in any one number place must be lined up properly.

   Example: 12.076 − 2.314 →   12.076
   $$\quad\quad\quad\quad\quad\quad - \underline{\quad 2.314}$$

   Write these decimal numbers in proper form for subtraction. (Do not solve.)

   a.   0.089 − 0.012

   b.   12.82 − 8.46

---

1. a.   0.103
   $$+ \underline{0.006}$$

   b.   13.123
   $$+ \underline{10.001}$$

2. a.   0.807
   $$+ \underline{0.116}$$
   $$\quad\;\; 0.923$$

   b.   1.76
   $$+ \underline{4.13}$$
   $$\quad\; 5.89$$

3. a.   0.02
   $$\quad\;\; 0.1006$$
   $$+ \underline{0.9}$$

   b.   1.8
   $$\quad\; 10.09$$
   $$+ \underline{3.001}$$

4. a.   1.01
   $$\quad\; 10.1$$
   $$+ \underline{12.006}$$
   $$\quad\; 23.116$$

   b.   22.007
   $$\quad\;\; 9.12$$
   $$+ \underline{31.8}$$
   $$\quad\; 62.927$$

6. Be sure to line up the digits properly if the numbers have a different number of digits.

   Example: 12.1 − 8.006 →   12.1

                                 − 8.006

   Rewrite these problems with the digits lined up properly. (Do not solve.)

   a.   18.1 − 2.04   b.   21.0007 − 2.1

7. Rewrite these problems and solve them.

   a.   10.98 − 8.24

   b.   11.067 − 3.03

   c.   22.008 − 12.1

8. Work the problems below.

   a.   0.204 + 0.137   b.   1.2 + 21.007

   c.   8.6 + 11.0016 + 16.07   d.   10.84 − 5.44

   e.   12.022 − 6.01   f.   6.007 − 2.09

5. a.   0.089
      − 0.012

   b.   12.82
      −   8.46

6. a.   18.1
      −   2.04

   b.   21.0007
      −   2.1

7. a.   10.98
      −   8.24
      ────────
         2.74

   b.   11.067
      −   3.03
      ────────
         8.037

   c.   22.008
      − 12.1
      ────────
         9.908

Part 6. Multiplying Decimal Numbers

1. Decimal numbers are multiplied in two steps: first multiply the numbers to get the numerical answers, then place the decimal point in the correct place in the answer.

Example:

Step 1.
```
    10.4
X    0.3
    ────
    312   ← numerical answer
```

Step 2. The number of decimal places in the answer is found by adding up the number of decimal places in the numbers multiplied.

```
    10.4   ←     (one decimal place)
X    0.3   ← +   (one decimal place)
    ────
    3.12   ←     (two decimal places)
```

Note that, when inserting the decimal point, the places are counted from right to left. Here are more examples.

```
     10.9            10.3             12.81
X     0.8       X       4        X      0.2
     ────            ────             ─────
     8.72            41.2             2.562
```

State the number of decimal places that would appear in the answer to each of these multiplication problems.

```
a.     0.6      b.     12.4      c.      9.3
   X  0.5          X      3         X  0.08
   ─────           ──────          ──────
```

---

2. Place the decimal point in the proper place in the answers below.

```
a.     1.8      b.    11.6      c.     22.4
   X  0.7          X   0.3         X  0.08
   ─────           ──────         ───────
    126             348             1792
```

---

3. Zeros must be inserted to the *left* of the answer when there are not enough digits for the required number of decimal places.

Example:
```
      0.2   ←     (one decimal place)        0.2
   X  0.3   ← +   (one decimal place)  =  X  0.3
      ───                                  ────
        6   ←     (two decimal places)      0.06
```

Here are more examples.

Step 1. (Multiply numbers)
```
a.      2.1          b.      3.3
    X  0.03              X  0.003
    ──────              ───────
        63                   99
```

Step 2. (Place decimal point)
```
a.      2.1          b.      3.3
    X  0.03              X  0.003
    ──────              ───────
     0.063              0.0099
```

8. a. 0.341
   b. 22.207
   c. 35.6716
   d. 5.40
   e. 6.012
   f. 3.917

1. a. two places
   b. one place
   c. three places

2. a. 1.26
   b. 3.48
   c. 1.792

Place the decimal points in the answers below.

a.   3.2
    X 0.02
    ――――――
        64

b.   4.2
    X 0.002
    ――――――
        84

c.   0.64
    X 0.02
    ――――――
       128

―――

4. Work the problems below, inserting the correct number of decimal places.

a.   4.1
    X 0.01
    ――――

b.   3.1
    X 0.003
    ――――――

c.   0.82
    X 0.03
    ――――

3. a. 0.064
   b. 0.0084
   c. 0.0128

―――

5. Work these problems, inserting the correct number of decimal places.

a.   24.6
    X 0.03
    ――――

b.   62.4
    X   2.4
    ――――

c.   8.36
    X 1.23
    ――――

4. a. 0.041
   b. 0.0093
   c. 0.0246

d.   1.12
    X   60
    ――――

e.   31.2
    X 2.06
    ――――

f.   0.234
    X   506
    ――――

―――

6. State the correct number of decimal places in the answers.

a.   2.46
    X 3.21
    ――――

b.   0.642
    X 0.005
    ――――――

5. a. 0.738
   b. 149.76
   c. 10.2828
   d. 67.20
   e. 64.272
   f. 118.404

Place the decimal point in the answers below.

c.   0.9
    X 0.4
    ――――
       36

d.   12.2
    X 0.04
    ――――
      488

e.   111
    X 0.12
    ――――
     1332

Work these problems, inserting the correct number of decimal places.

f.   4.7
    X 0.02
    ――――

g.   9.3
    X 0.003
    ――――――

h.   0.823
    X   0.21
    ――――

i.   73.1
    X   3.3
    ――――

j.   4.26
    X 2.41
    ――――

k.   83.2
    X 2.01
    ――――

l.   500
    X 0.021
    ――――――

m.   1.33
    X   50
    ――――

n.   0.836
    X 0.0101
    ――――――

Part 7. Dividing Decimal Numbers

1. There are two steps in dividing decimal numbers: first place the decimal point, and then find the numerical number.

   Example:

   Step 1. The decimal point in the answer is placed directly above the decimal point of the number being divided.

   $$6\overline{)4.8}\quad\text{— decimal points align}$$

   Step 2. $6\overline{)4.8} = 6\overline{)4.8}\quad \overset{0.8}{}\ \leftarrow\ \text{numerical answer}$

   Here are more examples.

   $$9\overline{)36.9} = 9\overline{)36.9}\ ^{4.1} \qquad\qquad 6\overline{)4.26} = 6\overline{)4.26}\ ^{0.71}$$

   Place the decimal point in these answers.

   a. $8\overline{)64.8}\ ^{8\ 1}$

   b. $7\overline{)2.17}\ ^{31}$

---

2. Work these problems, putting the decimal point in the right place.

   a. $9\overline{)72.9}$

   b. $9\overline{)4.59}$

   c. $12\overline{)1.44}$

---

3. Sometimes zeros must be inserted into the answer in order to obtain the correct number of decimal places.

   Example:

   Step 1. $12\overline{)0.36} = 12\overline{)0.36}\quad \leftarrow$ Place the decimal point.

   Step 2. $12\overline{)0.36} = 12\overline{)0.36}\ ^{0.03}\quad \leftarrow$ Find the numerical answer.

   Here are more examples.

   $$8\overline{)0.48} = 8\overline{)0.48}\ ^{0.06} \qquad\qquad 300\overline{)0.300} = 300\overline{)0.300}\ ^{0.001}$$

   Solve these problems, putting the decimal point in the right place.

   a. $12\overline{)0.48}$

   b. $600\overline{)0.1200}$

---

6. a. four places
   b. six places
   c. 0.36
   d. 0.488
   e. 13.32
   f. 0.094
   g. 0.0279
   h. 0.17283
   i. 241.23
   j. 10.2666
   k. 167.232
   l. 10.500
   m. 66.50
   n. 0.0084436

1. a. 8.1
   b. 0.31

2. a. 8.1
   b. 0.51
   c. 0.12

4. There must be no decimal point in the divisor (the number you are dividing by). If there is a decimal point in the divisor, the decimal point in *both* the divisor and the dividend (the number being divided) must be shifted to the right an equal number of places until the divisor is a whole number. Then perform the division in the usual manner.

Example: $1.5\overline{\smash{)}4.5} = 1\underset{\smallsmile}{5}\overline{\smash{)}4\underset{\smallsmile}{5}}$

dividend
divisor

Here are more examples.

$1.6\overline{\smash{)}0.32} = 16\overline{\smash{)}3.2}$ $0.07\overline{\smash{)}490} = 7\overline{\smash{)}49000}$

Rewrite these problems, showing the decimal-point shift. (Do not solve.)

a. $2.2\overline{\smash{)}0.44} =$

b. $0.006\overline{\smash{)}3.660} =$

c. $0.08\overline{\smash{)}0.0072} =$

5. Solve these division problems.

a. $0.9\overline{\smash{)}0.72} =$

b. $3.3\overline{\smash{)}9.9} =$

c. $0.08\overline{\smash{)}0.072} =$

d. $28.6\overline{\smash{)}343.2} =$

6. Work the problems below.

a. $0.06\overline{\smash{)}0.048}$ b. $12.1\overline{\smash{)}37.51}$

c. $0.008\overline{\smash{)}0.0736}$ d. $48\overline{\smash{)}0.216}$

3. a. 0.04
   b. 0.0002

4. a. $22\overline{\smash{)}4.4}$
   b. $6\overline{\smash{)}3660}$
   c. $8\overline{\smash{)}0.72}$

5. a. 0.8
   b. 3
   c. 0.9
   d. 12

Part 8. Using the Calculator

1. The hand calculator can be used to perform addition, subtraction, multiplication, and division. This addition:

$$8.65 + 19.7 + 5.43 =$$

can be performed as follows:

Enter	Press	Display
8.65	$+$	8.65
19.7	$+$	28.35
5.43	$=$	33.78

The answer is 33.78.

Use a calculator to perform these additions.

a.   15.09
     4.7
  + 35.9

b. 29.7 + 0.8 + 107.2 = _____

c. 9.3 + 0.06 + 3.4 = _____

d. 15.22 + 0.009 = _____

6. a. 0.8
   b. 3.1
   c. 9.2
   d. 0.0045

2. The subtraction at the right can be performed by using the calculator as follows:

    25.90
  − 1.56

Enter	Press	Display
25.90	$-$	25.90
1.56	$=$	24.34

The answer is 24.34.

Use a calculator to perform these subtractions.

a.   5285
  − 1760

b.   98.75
  − 4.895

c.   7.008
  − 2.017

1. a. 55.69
   b. 137.7
   c. 12.76
   d. 15.229

3. The multiplication at the right can be performed by using a hand calculator as follows:

    7.6
  X 907

Enter	Press	Display
7.6	$\times$	7.6
907	$=$	6893.2

The answer is 6893.2.

Use a hand calculator to perform the following multiplications.

a.   4.89
  X 12.2

b.   48
  X 17

c.   54.6
  X 879

d.   28.6
  X 45

2. a. 3525
   b. 93.855
   c. 4.991

4. The division at the right can be performed on the calculator as follows: $28\overline{)71.68}$

Enter	Press	Display
71.68	$\div$	71.68
28	$=$	2.56

The answer is 2.56.

Use a calculator to perform these divisions.

a. $1.4\overline{)53.2}$ b. $42\overline{)264.6}$ c. $24.8\overline{)414.16}$

3. a. 59.658
   b. 816
   c. 47,993.4
   d. 1287

5. Combined addition and subtraction operations can be performed with a calculator.

   Example: 23.79 + 0.54 − 6 =

Enter	Press	Display
23.79	$+$	23.79
0.54	$-$	24.33
6	$=$	18.33

The answer is 18.33.

Use a calculator to perform these combined operations.

a. 10.7 + 3.08 − 4.2 = _____   b. 24.06 − 9.7 + 8.3 = _____

c. 18.16 + 20.12 − 0.4 = _____

4. a. 38
   b. 6.3
   c. 16.7

6. Combined multiplication and division can be performed as well.

   Example: $\dfrac{28.7 \times 3.9}{4.2}$ =

Enter	Press	Display
28.7	$\times$	28.7
3.9	$\div$	111.93
4.2	$=$	26.65

The answer is 26.65.

Use a calculator to perform these combined operations.

a. $\dfrac{14.2 \times 9.2}{2.0}$ = _____   b. $\dfrac{12.8 \times 3.6}{7.2}$ = _____

c. $\dfrac{3.2 \times 10.17}{0.4}$ = _____   d. $\dfrac{1.8 \times 29.7}{9}$ = _____

5. a. 9.58
   b. 22.66
   c. 37.88

7. Here is another example of combined operations.

Example: $\dfrac{24}{4 \times 3}$ =

Enter	Press	Display
24	÷	24
4	÷	6
3	=	2

The answer is 2.

Notice that the number in the numerator is divided by each number in the denominator. Do not let the multiplication sign in the denominator confuse you. Enter the numbers and operations into the calculator just as you read them.

Example: $\dfrac{72}{9 \times 2}$ =

This reads: 72 divided by 9, divided by 2 equals.

Enter	Press	Display
72	÷	72
9	÷	8
2	=	4

The answer is 4.

Perform these combined operations.

a. $\dfrac{81}{3 \times 9}$ = _____

b. $\dfrac{144}{12 \times 3}$ = _____

c. $\dfrac{0.90}{0.10 \times 0.3}$ = _____

d. $\dfrac{1.20}{0.40 \times 1.5}$ = _____

8. Here is a more-complex combined-operation procedure.

Example: $\dfrac{8 \times 6}{2 \times 3}$ =

This reads: 8 times 6, divided by 2, divided by 3 equals.

Enter	Press	Display
8	X	8
6	÷	48
2	÷	24
3	=	8

The answer is 8.

6. a. 65.32
   b. 6.4
   c. 81.36
   d. 5.94

7. a. 3
   b. 4
   c. 30
   d. 2

Perform these combined operations.

a. $\dfrac{10 \times 12}{5 \times 6} = $ _____

b. $\dfrac{150 \times 8}{25 \times 4} = $ _____

c. $\dfrac{0.48 \times 4}{0.3 \times 2} = $ _____

d. $\dfrac{0.75 \times 80}{0.25 \times 100} = $ _____

---

9. Often you will find problems written in conversion-factor form.

Example: $200 \times \dfrac{300}{150}$

Multiply the first number (200), by the number in the numerator (300), then divide by the number in the denominator (150).

This reads: 200 times 300, divided by 150 equals.

Enter	Press	Display
200	X	200
300	÷	60,000
150	=	400

The answer is 400.

Perform these calculations.

a. $600 \times \dfrac{200}{300} = $ _____

b. $1200 \times \dfrac{50}{20} = $ _____

c. $0.8 \times \dfrac{20}{0.1} = $ _____

d. $0.16 \times \dfrac{0.9}{0.2} = $ _____

---

10. Here is a more-complex conversion-factor problem.

$$40 \times \dfrac{10}{20} \times \dfrac{30}{60} = $$

Multiply the first number (40) by each number in the numerator and divide by each number in the denominator.

This reads: 40 times 10, times 30, divided by 20, divided by 60 equals.

Enter	Press	Display
40	X	40
10	X	400
30	÷	12,000
20	÷	600
60	=	10

The answer is 10.

8. a. 4
   b. 12
   c. 3.2
   d. 2.4

9. a. 400
   b. 3000
   c. 160
   d. 0.72

Perform these calculations.

a. $50 \times \dfrac{10}{25} \times \dfrac{30}{20} =$ _____

b. $100 \times \dfrac{5}{20} \times \dfrac{2}{10} =$ _____

c. $1.5 \times \dfrac{30}{15} \times \dfrac{6.8}{3.4} =$ _____

d. $0.75 \times \dfrac{3}{2.5} \times \dfrac{0.5}{1.5} =$ _____

---

11. Perform these calculations on your calculator.

a. $7.92 + 18.69 + 20.13 =$ _____   b. $53.7 - 41.9 =$ _____

c. $3.5 \times 10.1 =$ _____   d. $0.99 - 0.3 =$ _____

e. $16.3 + 4.6 - 2.7 =$ _____   f. $\dfrac{14.22 \times 6.6}{3.3} =$ _____

g. $\dfrac{0.70}{0.1 \times 0.5} =$ _____   h. $\dfrac{0.56 \times 8.4}{0.70 \times 4.2} =$ _____

i. $0.32 \times \dfrac{0.34}{0.16} =$ _____   j. $8.8 \times \dfrac{18}{36} \times \dfrac{7.2}{2.4} =$ _____

10. a. 30
    b. 5
    c. 6
    d. 0.3

---

12. Answers to frame 11.

a. 46.74	b. 11.8	c. 35.35	d. 0.69	e. 18.2
f. 28.44	g. 14	h. 1.6	i. 0.68	j. 13.2

NAME _____

WHOLE NUMBERS AND DECIMALS
EVALUATION TEST 1
PART 1–PART 2

In the number 361,287,490, state what digit is in the indicated place.

1. thousands place

2. tens place

3. hundred-millions place

4. ones place

5. ten-millions place

6. hundreds place

7. hundred-thousands place

8. millions place

Write the number indicated by the word name.

9. four hundred thirty-two

10. seven hundred one

11. eight thousand, three hundred

12. seventy-six thousand, fifty

13. two hundred fourteen thousand, four hundred eight

14. five million, two hundred thousand

15. eighteen million, six hundred twelve

16. six hundred thirty-two million, nine hundred fourteen thousand, six hundred nine

1. _____

2. _____

3. _____

4. _____

5. _____

6. _____

7. _____

8. _____

9. _____

10. _____

11. _____

12. _____

13. _____

14. _____

15. _____

16. _____

WHOLE NUMBERS AND DECIMALS
EVALUATION TEST 2
PART 3–PART 4

In the number 0.00907, state what digit is in the indicated place.

1. hundredths place

2. ten-thousandths place

Write the word name of the decimal number.

3. 0.4

4. 0.17

5. 0.003

6. 0.142

Write the decimal number indicated by the word name.

7. nine tenths

8. four hundredths

9. eighty-eight thousandths

10. nine ten-thousandths

Name these decimal numbers.

11. 8.2

12. 6.03

13. 68.04

14. 3.0012

Write the decimal number indicated by the word name.

15. three point six

16. thirty-two point zero

17. seventy-two point zero three

18. twelve point three

1. _____

2. _____

3. _____

4. _____

_____

5. _____

_____

6. _____

_____

7. _____

8. _____

9. _____

10. _____

11. _____

12. _____

_____

13. _____

_____

14. _____

_____

15. _____

16. _____

17. _____

18. _____

WHOLE NUMBERS AND DECIMALS
EVALUATION TEST 3
PART 5—PART 7

Solve the problems below.

1. $0.306 + 0.253 =$

2. $2.2 + 31.0006 + 14.026 =$

3. $2.3 + 3.0012 =$

4. $12.88 - 6.42 =$

5. $8.003 - 4.09 =$

6. $\begin{array}{r} 2.7 \\ \times\ 0.03 \\ \hline \end{array}$

7. $\begin{array}{r} 8.2 \\ \times\ 0.004 \\ \hline \end{array}$

8. $\begin{array}{r} 0.876 \\ \times\ 2.01 \\ \hline \end{array}$

9. $0.8\overline{)0.64}$ 10. $2.2\overline{)4.4}$

11. $0.09\overline{)0.081}$ 12. $27.2\overline{)11.152}$

1. _____
2. _____
3. _____
4. _____
5. _____
6. _____
7. _____
8. _____
9. _____
10. _____
11. _____
12. _____

WHOLE NUMBERS AND DECIMALS
EVALUATION TEST 4
PART 8

Use a calculator to solve these problems.

1. $18.91 + 32.15 =$

2. $21.87 + 0.95 + 11.27 =$

3. $98.27 - 62.97 =$

4. $0.895 - 0.428 =$

5. $7.55 \times 10.98 =$

6. $0.87 \times 1.18 =$

7. $35.75 \div 14.3 =$

8. $0.9495 \div 0.211 =$

9. $28.6 - 14.2 + 18.7 =$

10. $0.987 + 1.154 - 0.243 =$

11. $\dfrac{44.44 \times 2.4}{9.6} =$

12. $\dfrac{0.139 \times 0.633}{0.211} =$

13. $\dfrac{48.4}{12.1 \times 1.25} =$

14. $\dfrac{0.99}{0.3 \times 1.1} =$

15. $\dfrac{1.8 \times 6.8}{0.9 \times 2} =$

16. $\dfrac{28.8 \times 4.8}{14.4 \times 3.2} =$

17. $9.7 \times \dfrac{1000}{100} =$

18. $400 \times \dfrac{300}{600} =$

19. $0.87 \times \dfrac{600}{300} \times \dfrac{546}{273} =$

20. $520 \times \dfrac{100}{1000} \times \dfrac{50}{500} =$

1. _____

2. _____

3. _____

4. _____

5. _____

6. _____

7. _____

8. _____

9. _____

10. _____

11. _____

12. _____

13. _____

14. _____

15. _____

16. _____

17. _____

18. _____

19. _____

20. _____

UNIT 2
Signed Numbers

In this unit you will learn operations with signed numbers.

Part 1. Definition of Signed Numbers

1. In chemistry we deal with numbers such as +3 and −4. These are called *signed numbers*. The + and − are called the *signs* of the numbers, and the 3 and 4 are called the *absolute value* of the numbers.

2. Here are two signed numbers: +6 and −6.

   a. Do the two numbers have the same sign? _____

   b. Do the two numbers have the same absolute value? _____

3. What is the absolute value of these numbers?

   a. −27   b. +82   c. −1157

2. a. no
   b. yes

Part 2. Adding Signed Numbers With Like Signs

1. If all the numbers to be added have the same sign, the sign of the sum is the same as that of the original numbers.

   Examples: (+2) + (+5) = +7
                 (−3) + (−6) = −9

3. a. 27
   b. 82
   c. 1157

Notice that the plus and minus signs have two different purposes in the addition problems above. One purpose is to indicate the sign of the number; the other is to indicate the mathematical operation to be performed.

Example: signs of numbers

                (+2) + (+5) = +7

                mathematical
                operation

In the equations below, what will the signs of the answers be?

   a. (+6) + (+3) = ?9   b. (−2) + (−4) = ?6

2. When all the numbers to be added have the same sign, the absolute value of the sum is obtained by adding the absolute values of the numbers.

Examples: (+4) + (+3) = +7   (−2) + (−7) = −9

Give the answers to the problems below.

a.   (+3) + (+4) = _____   b.   (−6) + (−1) = _____

c.   (+1) + (+19) = _____   d.   (−18) + (−12) = _____

Part 3. Adding Signed Numbers With Unlike Signs

1. If two numbers to be added have different signs, the sign of the sum is the same as the sign of the number with the larger absolute value.

Examples: (+8) + (−4) = +4

      sign of number
      with larger
      absolute value

(−10) + (+5) = −5

      sign of number
      with larger
      absolute value

What will be the sign of the answers in the problems below?

a.   (−8) + (+2) = ?6   b.   (−4) + (+6) = ?2

2. When two numbers to be added have different signs, the absolute value of the sum is obtained by *subtracting* the smaller absolute value from the larger absolute value.

Examples: (−8) + (+2) = ?   8 − 2 = 6   (Subtract the smaller absolute value from the larger absolute value.)

      (−8) + (+2) = −6   (Use the sign of the larger.)

    (−4) + (+6) = ?   6 − 4 = 2
    (−4) + (+6) = +2

Give the answers to the problems below, indicating the correct signs and absolute values.

a.   (+6) + (−3) = _____   b.   (+3) + (−7) = _____

c.   (−8) + (+2) = _____   d.   (−4) + (+5) = _____

3. Work the set of problems below.

a.   (−7) + (−2) = _____   b.   (+2) + (−5) = _____

c.   (−9) + (+10) = _____   d.   (+6) + (−5) = _____

e.   (−8) + (+5) = _____   f.   (−3) + (−2) = _____

g.   (+13) + (−33) = _____   h.   (−12) + (−13) = _____

i.   (+24) + (+20) = _____   j.   (−19) + (+49) = _____

Part 4. Subtracting Signed Numbers

1. Subtraction of two signed numbers is accomplished by changing the problem to addition. Change the subtraction sign to an addition sign and change the sign of the last number in the problem. The rules of addition then apply.

    Examples: a. $(-2) - (-4) = (-2) + (+4) =$

      Change signs.

      b. $(-6) - (+2) = (-6) + (-2) =$

      Change signs.

    Change the following subtraction problems to addition.

    a. $(+3) - (+9)$ _____

    b. $(+7) - (-7)$ _____

    c. $(-3) - (+4)$ _____

    d. $(-8) - (-7)$ _____

2. We can now go through the steps of changing subtraction to addition and using addition rules to solve the problem.

    Examples: a. $(-9) - (-3)$

    Change to addition.   → $(-9) + (+3)$

    Use addition rules.   → $(-9) + (+3) = -6$

                    b. $(+6) - (+3)$

    Change to addition.   → $(+6) + (-3)$

    Use addition rules.   → $(+6) + (-3) = +3$

    Solve these subtraction problems using the steps described in frames 1 and 2.

    a. $(+4) - (+8) =$ _____   b. $(+6) - (-2) =$ _____

    c. $(-3) - (+8) =$ _____   d. $(-8) - (-9) =$ _____

3. a. −9
   b. −3
   c. +1
   d. +1
   e. −3
   f. −5
   g. −20
   h. −25
   i. +44
   j. +30

1. a. $(+3) + (-9)$
   b. $(+7) + (+7)$
   c. $(-3) + (-4)$
   d. $(-8) + (+7)$

3. Decimal numbers also can be signed numbers.

Examples: $(-0.3) + (-1.2) = -1.5$
$(-4.6) + (+5.2) = +0.6$
$(-8.6) - (+3.5) = -12.1$

Solve these problems.

a. $(+0.9) + (+1.8) =$ _____

b. $(-2.64) + (-3.186) =$ _____

c. $(-1.874) + (+2.965) =$ _____

d. $(+8.62) + (-4.65) =$ _____

e. $(-1.987) - (-2.6) =$ _____

f. $(+6.94) - (-8.9) =$ _____

4. Here is a summary of the rules for adding and subtracting signed numbers.

Adding
(The numbers to be added have the same sign.)
The sign of the sum is the same as that of the numbers added.
The absolute value of the sum is obtained by adding the absolute values of the numbers.

(The numbers to be added have unlike signs.)
The sign of the sum is the same as the sign of the number with the larger absolute value.
The absolute value of the sum is obtained by subtracting the smaller absolute value from the larger absolute value of the numbers being added.

Subtracting
Change the problem to addition and use addition rules.

5. Do the set of problems below.

a. $(+6) - (+3) =$ _____

b. $(-6) - (+2) =$ _____

c. $(+2) - (+5) =$ _____

d. $(+3) - (-8) =$ _____

e. $(-3) - (-2) =$ _____

f. $(-5) - (+6) =$ _____

g. $(+6) - (-4) =$ _____

h. $(-8) - (-5) =$ _____

i. $(-3) - (-5) =$ _____

j. $(+9) - (+4) =$ _____

k. $(-7.94) + (-8.96) =$ _____

l. $(-18.6) + (+5.7) =$ _____

m. $(-1.74) - (-3.6) =$ _____

n. $(+9.51) - (+7.2) =$ _____

2. a. -4
b. $+8$
c. -11
d. $+1$

3. a. 2.7
b. -5.826
c. 1.091
d. 3.97
e. 0.613
f. 15.84

Part 5. Using the Calculator to Perform Operations With Signed Numbers

1. The hand calculator can be used to perform operations with signed numbers. The $\boxed{+/-}$ key is used for these operations.

   The addition at the right can be done using a hand calculator. $(-5) + 4 =$

Enter	Press	Display
5	$\boxed{+/-}$ $\boxed{+}$	−5
4	$\boxed{=}$	−1

   Therefore $(-5) + 4 = -1$.

   Use a hand calculator to perform the following additions.

   a. $(-9) + 7 =$ _____   b. $(-58) + 19 =$ _____

2. The addition at the right is performed by using a hand calculator. $(+9) + (-7) =$

Enter	Press	Display
9	$\boxed{+}$	9
7	$\boxed{+/-}$ $\boxed{=}$	2

   Therefore $(+9) + (-7) = 2$.

   Use a hand calculator to perform the following additions.

   a. $7 + (-8) =$ _____   b. $95 + (-40) =$ _____

3. The addition at the right is performed by using a hand calculator. $(-8) + (-7) =$

Enter	Press	Display
8	$\boxed{+/-}$ $\boxed{+}$	−8
7	$\boxed{+/-}$ $\boxed{=}$	−15

   Therefore $(-8) + (-7) = -15$.

   Use a hand calculator to perform the following additions.

   a. $(-9) + (-4) =$ _____   b. $(-87) + (-12) =$ _____

5. a. +3
   b. −8
   c. −3
   d. +11
   e. −1
   f. −11
   g. +10
   h. −3
   i. +2
   j. +5
   k. −16.90
   l. −12.9
   m. 1.86
   n. 2.31

1. a. −2
   b. −39

2. a. −1
   b. +55

4. The subtraction at the right is peformed by using a hand calculator.   (5) – (–7) =   3. a. – 13
b. – 99

Enter	Press	Display
5	$\boxed{-}$	5
7	$\boxed{+/-}$ $\boxed{=}$	12

Therefore (5) – (–7) = 12.

The $\boxed{+/-}$ key is used to enter a negative number.

Use a hand calculator to perform the following subtractions.

a. (7) – (–2) = _____   b. (–8) – (–5) = _____

c. (–9) – (–4) = _____

5. The multiplication at the right is performed by using a hand calculator.   (–5) (8) =   4. a. 9
b. –3
c. – 5

Enter	Press	Display
5	$\boxed{+/-}$ $\boxed{X}$	–5
8	$\boxed{=}$	–40

Therefore (–5) (8) = –40.

Use a hand calculator to perform the following multiplications.

a. (– 9) (8) = _____   b. (–97) (18) = _____

6. The multiplication at the right is performed by using a hand calculator.   (9) (–7) =   5. a. –72
b. –1746

Enter	Press	Display
9	$\boxed{X}$	9
7	$\boxed{+/-}$ $\boxed{=}$	–63

Therefore (9) (–7) = –63.

Use a hand calculator to perform the following multiplications.

a. (7) (–8) = _____   b. (95) (–42) = _____

7. The multiplication at the right is performed by using a hand calculator.   (–7) (–6) =   6. a. –56
b. –3990

Enter	Press	Display
7	$\boxed{+/-}$ $\boxed{X}$	–7
6	$\boxed{+/-}$ $\boxed{=}$	42

Therefore (–7) (–6) = +42.

Use a hand calculator to perform the following multiplications.

a. (–8) (–4) = _____   b. (–31) (–47) = _____

8. The same procedures are used to perform division with a hand calculator. The [+/−] key is used to enter a negative number.

The division at the right is performed by using a hand calculator.

$$\frac{-48}{-6}$$

Enter	Press	Display
48	[+/−] [÷]	−48
6	[+/−] [=]	8

Therefore $\frac{-48}{-6} = 8$.

Use a hand calculator to perform the following divisions.

a. $\frac{-64}{8} =$ _____

b. $\frac{-400}{-20} =$ _____

c. $\frac{180}{-45} =$ _____

d. $\frac{-3200}{-40} =$ _____

9. Use a hand calculator to perform the following operations.

a. $(-3.5) + 4.2 =$ _____

b. $2.97 + (-3.12) =$ _____

c. $(-4.6) + (-0.3) =$ _____

d. $(-0.87) + (-1.12) =$ _____

e. $(-9.5) - (6.5) =$ _____

f. $2.13 - (-4.8) =$ _____

g. $(-10.28) - (-4.61) =$ _____

h. $(-3.2)(4.1) =$ _____

i. $(9.8)(-1.7) =$ _____

j. $(-1.3)(-4.9) =$ _____

k. $\frac{-50.43}{12.3} =$ _____

l. $\frac{-31.04}{-9.7} =$ _____

10. Answers to frame 9.

a. 0.7 b. −0.15 c. −4.9 d. −1.99 e. −16 f. 6.93
g. −5.67 h. −13.12 i. −16.66 j. 6.37 k. −4.1 l. 3.2

7. a. +32
   b. +1457

8. a. −8
   b. +20
   c. −4
   d. +80

NAME _____

SIGNED NUMBERS
EVALUATION TEST
PART 1-PART 5

Give the answers to the addition problems below, with the correct signs and absolute values.

1. $(-6) + (-3) =$

2. $(+3) + (-4) =$

3. $(-8) + (+9) =$

4. $(+6) + (-4) =$

5. $(-8) + (+6) =$

6. $(-2) + (-3) =$

7. $(+13) + (-21) =$

8. $(-21) + (-13) =$

9. $(+24) + (+20) =$

10. $(-19) + (+49) =$

Give the answers to the subtraction problems below, with the correct signs and absolute values. Show the step of changing subtraction to addition.

11. $(+5) - (+2) =$ 12. $(-4) - (+7) =$

13. $(-4) - (+3) =$ 14. $(+6) - (-3) =$

15. $(+2) - (+6) =$ 16. $(-8) - (-6) =$

17. $(+3) - (-7) =$ 18. $(-3) - (-8) =$

19. $(-4) - (-1) =$ 20. $(+9) - (+3) =$

21. $(-3.8) - (+2.6) =$ 22. $(+4.75) - (-2.13) =$

1. _____

2. _____

3. _____

4. _____

5. _____

6. _____

7. _____

8. _____

9. _____

10. _____

11. _____

12. _____

13. _____

14. _____

15. _____

16. _____

17. _____

18. _____

19. _____

20. _____

21. _____

22. _____

UNIT 3
Powers of Ten and Logarithms

In this unit you will have to do simple calculations with numbers in power-of-ten form. In chemistry textbooks numbers in power-of-ten form are also referred to as numbers in exponential notation or scientific notation. You will learn how to write numbers in power-of-ten form, how to multiply them, and how to divide them. You will also learn what a logarithm is and how to find logarithms using a calculator.

Part 1. Writing Numbers in Powers-of-Ten Form

1. The power-of-ten form of writing a number is a shorthand method of writing a very large or very small number.

   Examples: a. $1000 = 1 \times 10 \times 10 \times 10$
   or
   $1000 = 1 \times 10^3$

   b. $10,000 = 1 \times 10 \times 10 \times 10 \times 10$
   or
   $10,000 = 1 \times 10^4$

   In the number 10^4 the 10 is called the *base* and the 4 is called the *exponent*. Write the power-of-ten form of the numbers below.

   a. $100 =$ _____   b. $100,000 =$ _____

2. There is a faster method of determining what a number will be in power-of-ten form. Move the decimal point so that there is only one digit to the left of it. Count the number of spaces you have moved the decimal point and this becomes your exponent on your base 10.

   Examples: a.   There is an implied decimal point at the end of a whole number. For example, (100) can be written (100.). To put this in power-of-ten form, move the decimal point so that only one digit remains to the left of it.

   $$100. \rightarrow 1.00$$

   We have moved the decimal point two places to the left so the exponent becomes 2 on the base 10, or 1.00×10^2. Final zeros may be dropped if the number is not a measurement. (In a later unit you will learn rules for dropping zeros.) The number 1.00×10^2 is usually written as 1×10^2.

1. a. 1×10^2
   b. 1×10^5

39

b. (1000) can be written (1000.). Move the decimal point so that only one digit remains to the left.

$$1000 \rightarrow 1.000$$

The number of places the decimal point has been moved—3—becomes the exponent on the base: 1×10^3.

c. (10,000) is (10,000.).

$$10,000 \rightarrow 1.0000$$

The number is 1×10^4.

Write the power-of-ten form of these numbers, dropping the zeros after the decimal point.

a. 100 _____

b. 100,000 _____

c. 100,000,000 _____

d. 10,000 _____

e. 1000 _____

f. 1,000,000 _____

3. The number 1 is a special case.

Example: (1) can be written (1.). Because we must leave one digit to the left of the decimal point, we cannot move the decimal point at all. Therefore, $1 = 1 \times 10^0$.

2. a. 1×10^2
   b. 1×10^5
   c. 1×10^8
   d. 1×10^4
   e. 1×10^3
   f. 1×10^6

4. So far we have looked only at numbers that begin with the digit 1. Numbers beginning with any digit or set of digits, however, can be placed in power-of-ten form.

Examples: a. (200) is (200.). The decimal point is moved so that only one digit remains to the left of it: (200.) becomes (2.00). The number of places the decimal has been moved becomes the exponent on the base, and the number is written in the following manner: 2.00×10^2.

b. (9000) is (9000.).

$$9000. \rightarrow 9.000$$

The number is 9×10^3.

5. A number to be written in power-of-ten form also can have more than one nonzero digit.

Example: (20,200) is (20,200.).

Step 1. (20,200.) becomes 2.0200
Step 2. 2.0200 becomes 2.02×10^4.

Write these numbers in power-of-ten form, dropping the zeros at the end of the number.

a. 6960 _____

b. 123,000 _____

c. 80,200,000 _____

6. We can also write decimal numbers in power-of-ten form. The same rules apply as in whole numbers: Move the decimal point so that only one digit remains to the left of it. Count the number of spaces you moved the decimal point and this becomes your exponent on your base 10. With decimal numbers, however, add a negative sign to the exponent to show that it is a decimal number in power-of-ten form.

Examples:　a.　(0.01) becomes (01.).
The decimal point was moved two places to the right so the exponent becomes −2. The minus sign on the exponent tells us that this is the power-of-ten form of a decimal number. The zero before the digit is dropped. The number in power-of-ten form is 1×10^{-2}.

　　　　　b.　(0.0001) becomes (0001.).
The exponent is −4 and the number is 1×10^{-4}. The zeros before the one are dropped.

　　　　　c.　(0.000001) becomes (000001.).
The number is 1×10^{-6}.

Put these numbers in power-of-ten form.

a.　0.0001 _____

b.　0.00001 _____

c.　0.0000001 _____

5. a. 6.96×10^3
　 b. 1.23×10^5
　 c. 8.02×10^7

---

7. Any decimal number can be put in power-of-ten form.

Examples:　a.　(0.012) becomes (01.2).
The decimal point was moved two places to the right so the exponent becomes −2. (The minus sign on the exponent tells us that this is the power-of-ten form of a decimal number.) The zero before the digit is dropped, and the number in power-of-ten form becomes 1.2×10^{-2}.

　　　　　b.　(0.00062) becomes (0006.2).
The exponent is −4 and the number is 6.2×10^{-4}. The zeros before the 6 are dropped.

　　　　　c.　(0.0000302) becomes (00003.02).
The number is 3.02×10^{-5}.

Put these numbers in power-of-ten form.

a.　0.0046 = _____

b.　0.000983 = _____

c.　0.00000402 = _____

6. a. 1×10^{-4}
　 b. 1×10^{-5}
　 c. 1×10^{-7}

---

8. Put the numbers in power-of-ten form.

a.　1000 _____　　b.　1,000,000 _____

c.　1 _____　　d.　909,000 _____

e.　67,000,000 _____　　f.　0.001 _____

g.　0.1 _____　　h.　0.00028 _____

i.　0.00306 _____　　j.　0.000000614 _____

7. a. 4.6×10^{-3}
　 b. 9.83×10^{-4}
　 c. 4.02×10^{-6}

Part 2. Multiplying Numbers in Powers-of-Ten Form

1. Multiplying numbers in power-of-ten form involves two steps.

   Step 1.   First multiply the two initial digits.

   $$(1 \times 10^2) \times (1 \times 10^3) = ?$$

   └─ Multiply. ─┘

   $$1 \times 1 = 1$$

   Step 2.   Multiply the numbers with exponents by *adding* the exponents. Put the signs in front of the exponents when adding.

   $$(1 \times 10^2) \times (1 \times 10^3) = ?$$

   Multiply by adding exponents.

   $$10^2 \times 10^3 = 10^{(+2) + (+3)}$$
   $$10^2 \times 10^3 = 10^5$$

   Now combine the answers from both steps.

   $$(1 \times 10^2) \times (1 \times 10^3) = 1 \times 10^5$$

   Multiply these power-of-ten numbers.

   a. $(1 \times 10^2) \times (3 \times 10^4) = ?$

   Step 1 _____

   Step 2 _____

   combined answer: _____

   b. $(2 \times 10^3) \times (4 \times 10^5) = ?$

   Step 1 _____

   Step 2 _____

   combined answer: _____

2. Multiply these power-of-ten numbers.

   a. $(2 \times 10^2) \times (4 \times 10^3) =$ _____

   _____

   _____

   b. $(3 \times 10^3) \times (3 \times 10^3) =$ _____

   _____

   _____

   c. $(4 \times 10^6) \times (1 \times 10^8) =$ _____

   _____

   _____

   d. $(3 \times 10^2) \times (1 \times 10^0) =$ _____

   _____

   _____

8. a. 1×10^3
   b. 1×10^6
   c. 1×10^0
   d. 9.09×10^5
   e. 6.7×10^7
   f. 1×10^{-3}
   g. 1×10^{-1}
   h. 2.8×10^{-4}
   i. 3.06×10^{-3}
   j. 6.14×10^{-7}

1. a. Step 1:
      $1 \times 3 = 3$
      Step 2:
      $10^2 \times 10^4 = 10^6$
      Answer: 3×10^6
   b. Step 1:
      $2 \times 4 = 8$
      Step 2:
      $10^3 \times 10^5 = 10^8$
      Answer: 8×10^8

3. We can also multiply power-of-ten numbers that have negative exponents.

Examples: a. $(1 \times 10^3) \times (2 \times 10^{-2}) = ?$

Step 1. $1 \times 2 = 2$
Step 2.   Add the exponents on the base 10, putting in the signs of the numbers.

$$10^3 \times 10^{-2} = 10^{(+3)+(-2)}$$
$$10^3 \times 10^{-2} = 10^1$$
combined answer: 2×10^1

b. $(2 \times 10^{-3}) \times (3 \times 10^{-3}) = ?$

Step 1. $2 \times 3 = 6$
Step 2. $10^{-3} \times 10^{-3} = 10^{(-3)+(-3)}$
$$10^{-3} \times 10^{-3} = 10^{-6}$$
combined answer: 6×10^{-6}

Remember: When multiplying numbers with exponents, *add* the exponents.

Multiply these numbers.

a. $(1 \times 10^3) \times (2 \times 10^1) = $ _____

_____

_____

b. $(2 \times 10^{-1}) \times (3 \times 10^{-2}) = $ _____

_____

_____

c. $(2 \times 10^2) \times (4 \times 10^{-1}) = $ _____

_____

_____

d. $(4 \times 10^{-1}) \times (2 \times 10^5) = $ _____

_____

_____

e. $(3 \times 10^{-3}) \times (3 \times 10^{-3}) = $ _____

_____

_____

f. $(4 \times 10^{-1}) \times (2 \times 10^{+1}) = $ _____

_____

_____

4. Sometimes when we multiply power-of-ten numbers, the answers are no longer in power-of-ten form.

Examples: a. $(8 \times 10^2) \times (2 \times 10^3) = \ 16 \times 10^5$
In power-of-ten form there can be only one digit to the left of the decimal point. We can change such answers back to power-of-ten form by moving the decimal point so that only one digit remains to the left of it and adding the number of spaces we moved the decimal point to the exponent.

2. a. 8×10^5
b. 9×10^6
c. 4×10^{14}
d. 3×10^2

3. a. 2×10^4
b. 6×10^{-3}
c. 8×10^1
d. 8×10^4
e. 9×10^{-6}
f. 8×10^0

Step 1. 16 $\times$ 10^5 becomes 1.6 $\times$ 10^5.
Step 2. We moved the decimal point one place, so we add 1 to the exponent:
1.6 $\times$ 10$^{(+5)+(+1)}$ = 1.6 $\times$ 10^6.

b. (9 $\times$ 10^{-5}) $\times$ (9 $\times$ 10^3) = 81 $\times$ 10^{-2}.

Step 1. 81 $\times$ 10^{-2} becomes 8.1 $\times$ 10^{-2}.
Step 2. Adding one to the exponent, 8.1 $\times$ 10$^{(-2)+(+1)}$ = 8.1 $\times$ 10^{-1}.

Multiply these numbers and put them in correct power-of-ten form. (Whenever two sets of parentheses containing numbers are adjoining and no operational sign is between them, it indicates that the two sets of numbers are to be multiplied.)

a. (9 $\times$ 10^2)(2 $\times$ 10^1) = _____

_____

_____

b. (8 $\times$ 10^7)(3 $\times$ 10^{-4}) = _____

_____

_____

c. (5.9 $\times$ 10^{-8})(3.1 $\times$ 10^4) = _____

_____

_____

d. (9 $\times$ 10^{-3})(2 $\times$ 10^{-3}) = _____

_____

_____

e. (4 $\times$ 10^{-1})(3 $\times$ 10^1) = _____

_____

_____

f. (6.2 $\times$ 10^{-1})(2.7 $\times$ 10^{-1}) = _____

_____

_____

Part 3. Dividing Numbers in Powers-of-Ten Form

1. Dividing numbers in power-of-ten form also occurs in two steps.

Example:

Step 1. Divide the initial digit in the numerator by the initial digit in the denominator.

$$\frac{\boxed{4} \times 10^4}{\boxed{2} \times 10^2} \quad \text{or} \quad \frac{4}{2} = 2$$

↑
Divide.

4. a. 1.8 $\times$ 10^4
   b. 2.4 $\times$ 10^4
   c. 1.829 $\times$ 10^{-3}
   d. 1.8 $\times$ 10^{-5}
   e. 1.2 $\times$ 10^1
   f. 1.674 $\times$ 10^{-1}

Step 2. Divide the numbers with exponents by *subtracting* the exponents in the denominator from the exponent in the numerator.

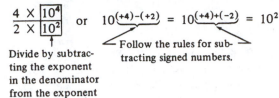

$$\frac{4 \times 10^4}{2 \times 10^2}$$ or $10^{(+4)-(+2)} = 10^{(+4)+(-2)} = 10^2$

Divide by subtracting the exponent in the denominator from the exponent in the numerator.

Follow the rules for subtracting signed numbers.

Now combine the answers from both steps.

$$\frac{4 \times 10^4}{2 \times 10^2} = 2 \times 10^2$$

Here is another example.

Step 1. Divide digits: $\dfrac{9 \times 10^5}{3 \times 10^{-2}} \rightarrow \dfrac{9}{3} = 3$

Step 2. Subtract exponents: $\dfrac{10^5}{10^{-2}} \rightarrow 10^{(+5)-(-2)} = 10^{(+5)+(+2)} = 10^7$

Combine steps. $\dfrac{9 \times 10^5}{3 \times 10^{-2}} = 3 \times 10^7$

Solve these problems.

a. $\dfrac{16 \times 10^5}{4 \times 10^3} =$ _____

_____

_____

b. $\dfrac{18 \times 10^5}{3 \times 10^{-2}} =$ _____

_____

_____

c. $\dfrac{21 \times 10^{-8}}{7 \times 10^2} =$ _____

_____

_____

d. $\dfrac{36 \times 10^{-9}}{9 \times 10^{-2}} =$ _____

_____

_____

2. Sometimes when we divide power-of-ten numbers, the answers are no longer in power-of-ten form.

Example: $\dfrac{2 \times 10^4}{4 \times 10^2} = 0.5 \times 10^2$

In power-of-ten form, there must be one digit to the left of the decimal point. So 0.5×10^2 becomes $5. \times 10^2$. Because we have moved the decimal point one place and we are dealing with a decimal number, we add a -1 to the exponent. We will do this in steps.

Step 1. Shift the decimal point: 0.5×10^2 becomes $5. \times 10^2$.
Step 2. Add a -1 to the exponent: $5 \times 10^{(+2)+(-1)} = 5 \times 10^1$.

Here is another example. $\dfrac{3 \times 10^5}{4 \times 10^7} = 0.75 \times 10^{-2}$

Step 1. Shift the decimal point: 0.75×10^{-2} becomes 7.5×10^{-2}.
Step 2. Add a -1 to the exponent: $7.5 \times 10^{(-2)+(-1)} = 7.5 \times 10^{-3}$.

Solve these problems.

a. $\dfrac{2 \times 10^5}{4 \times 10^3} =$

b. $\dfrac{6 \times 10^5}{8 \times 10^8} =$

3. Here is a summary of the rules for using numbers in power-of-ten form.

Writing
Step 1. Move the decimal point so that there is only one whole-number digit to the left of it.
Step 2. Count the number of spaces you moved the decimal point and this number becomes your exponent on the base 10.
Step 3. If the original number is a whole number, the exponent is positive. If the original number is a decimal number, the exponent is negative.

Multiplying
Step 1. Multiply the two initial digits.
Step 2. Add the exponents on the base 10.
Step 3. Adjust the answer, if necessary, to return it to the proper power-of-ten form.

Dividing
Step 1. Divide the two initial digits.
Step 2. Subtract the exponent on the base 10 in the denominator from the exponent on the base 10 in the numerator.
Step 3. Adjust the answer, if necessary, to return it to the proper power-of-ten form.

Remember: all answers must be in correct power-of-ten form.

1. a. 4×10^2
   b. 6×10^7
   c. 3×10^{-10}
   d. 4×10^{-7}

2. a. 5×10^1
   b. 7.5×10^{-4}

4. Put these numbers in power-of-ten form.

a. 100,000 _____   b. 1 _____

c. 317,000 _____   d. 0.001 _____

e. 0.1 _____   f. 0.00043 _____

Find the correct answers for the problems below.

g. $(2 \times 10^{-1})(3 \times 10^{-2}) =$

h. $(4 \times 10^{-4})(2 \times 10^{5}) =$

i. $(9 \times 10^{2})(2 \times 10^{3}) =$

j. $(4 \times 10^{-3})(5 \times 10^{-2}) =$

k. $\dfrac{16 \times 10^{5}}{4 \times 10^{1}} =$

l. $\dfrac{12 \times 10^{-5}}{4 \times 10^{2}} =$

m. $\dfrac{4 \times 10^{-1}}{8 \times 10^{2}} =$

n. $\dfrac{1 \times 10^{4}}{4 \times 10^{-2}} =$

Part 4. Logarithms

1. The logarithm is needed for a variety of problems in chemistry. Logarithms of numbers that are multiples or subdivisions of ten are easy to find. Any number can be written as 10 with some exponent. The exponent is called the logarithm or log of the number to base 10.

> The logarithm of 1×10^{2} is 2   (2 is the exponent on the base 10)
> The logarithm of 1×10^{-3} is -3 (-3 is the exponent on the base 10)

What is the logarithm of these numbers?

a. 1×10^{5} _____   b. 1×10^{4} _____

c. 1×10^{-2} _____   d. 1×10^{-5} _____

2. A number not in the power-of-ten form must be placed in that form before the logarithm can be obtained.

> The logarithm of 1000 = logarithm of $1 \times 10^{3} = 3$
> The logarithm of 0.01 = logarithm of $1 \times 10^{-2} = -2$

What is the logarithm of these numbers?

a. 10,000 _____   b. 100 _____

c. 0.001 _____   d. 0.00001 _____

4. a. 1×10^{5}
   b. 1×10^{0}
   c. 3.17×10^{5}
   d. 1×10^{-3}
   e. 1×10^{-1}
   f. 4.3×10^{-4}
   g. 6×10^{-3}
   h. 8×10^{1}
   i. 1.8×10^{6}
   j. 2.0×10^{-4}
   k. 4×10^{4}
   l. 3×10^{-7}
   m. 5×10^{-4}
   n. 2.5×10^{5}

1. a. 5
   b. 4
   c. -2
   d. -5

3. The notation used to represent a logarithm is:

$$y = \log_{10} X$$

This reads y is the logarithm to the base 10 of the number X. Another way of stating this is that y is the exponent that will be placed on the base 10 to equal the number X.

Some examples will illustrate this.

If $X = 100$, what is the logarithm of X?

Step 1. Substitute into the log formula.

$$y = \log_{10} 100$$

Step 2. Write X in scientific notation.

$$y = \log_{10} (1 \times 10^2)$$

Step 3. Solve for y.

$$y = 2.$$

y is the exponent used on the base 10 to equal X.

$$10^y = X$$
$$10^2 = 100$$

If $X = 0.001$, what is the logarithm of X?

Step 1. Substitute into the log formula.

$$y = \log_{10} 0.001$$

Step 2. Write X in scientific notation.

$$y = \log_{10} (1 \times 10^{-3})$$

Step 3. Solve for y.

$$y = -3$$

Check your answer by using y as the exponent on the base 10.

$$10^y = X$$
$$10^{-3} = 0.001$$

Solve for y in the following log equations.

a. $y = \log_{10} 1000$ _____

b. $y = \log_{10} 0.01$ _____

c. $y = \log_{10} 10,000$ _____

d. $y = \log_{10} 0.0001$ _____

4. Logarithms are very easy to find with a scientific calculator.

Example: Find the log of 8600.

Enter	Press	Display
8600	LOG	3.9345

$y = \log_{10} 8600 = 3.9345$ (rounded to four decimal places)

Example: Find the log of 0.000375.

Enter	Press	Display
0.000375	LOG	−3.4260

$y = \log_{10} 0.000375 = -3.4260$

Find the logarithm of these numbers with a calculator.

a. 3125 _____

b. 487,000 _____

c. 0.00861 _____

d. 0.00327 _____

2. a. 4
   b. 2
   c. −3
   d. −5

3. a. 3
   b. −2
   c. 4
   d. −4

4. a. 3.4949
   b. 5.6875
   c. −2.0650
   d. −2.4855

NAME _____

POWERS OF TEN
EVALUATION TEST
PART 1—PART 3

Put these numbers in power-of-ten form.

1.	10,000	2.	1
3.	806,000	4.	219,000
5.	28,000,000	6.	0.0001
7.	0.1	8.	0.00038
9.	0.0101	10.	0.00000064

Give correct answers for the problems below.

11. $(1 \times 10^3)(2 \times 10^4) =$ 16. $(5 \times 10^{-1})(5 \times 10^{-2}) =$

12. $(8 \times 10^2)(2 \times 10^1) =$ 17. $\dfrac{16 \times 10^3}{4 \times 10^1} =$

13. $(2 \times 10^{-2})(4 \times 10^{-3}) =$ 18. $\dfrac{15 \times 10^{-5}}{5 \times 10^{-3}} =$

14. $(6 \times 10^7)(3 \times 10^{-4}) =$ 19. $\dfrac{4 \times 10^{-1}}{5 \times 10^2} =$

15. $(4 \times 10^{-5})(2 \times 10^3) =$ 20. $\dfrac{1 \times 10^3}{5 \times 10^{-2}} =$

Use a calculator to find the logarithms of these numbers.

21. 2.77   22. 5628

23. 0.0007   24. 0.00894

1. _____
2. _____
3. _____
4. _____
5. _____
6. _____
7. _____
8. _____
9. _____
10. _____
11. _____
12. _____
13. _____
14. _____
15. _____
16. _____
17. _____
18. _____
19. _____
20. _____
21. _____
22. _____
23. _____
24. _____

UNIT 4
Measurement Concepts and Dimensional Analysis

In this unit you will learn to use simple measurement concepts. You will learn the rules for rounding off numbers, learn to apply the concept of precision of measurement, and learn the rules for determining significant digits. You will learn the process of dimensional analysis.

Part 1. Introduction

In a science course you often measure various quantities. It is important that the number representing the measurement be a reasonable number. In this unit you will learn rules to help you keep your answers reasonable. These rules will pertain *only* to numbers that represent measurements.

4. a. 3.4949
   b. 5.6875
   c. −2.0650
   d. −2.4855

Part 2. Rounding Off Numbers

1. When a number is rounded off to a given place, the digit immediately to the right of that place tells us what to do. If the digit to the immediate right is 5, 6, 7, 8, or 9, we add 1 to the place we are rounding to. If the digit to the immediate right is 0, 1, 2, 3, or 4, we simply eliminate that digit.

   Examples: a.   The number 2.489 is to be rounded off to the tenths place.

   ```
              ┌──── tenths place
              ↓
        2.489
              ↑
              └──── The digit in the hundredths place (8) is
                    5 or more, so we add 1 to the tenths place.
   ```

   2.489 (rounded off to the tenths place) = 2.5

   b.   The number 0.0925 is to be rounded off to the hundredths place.

   ```
              ┌──── hundredths place
              ↓
        0.0925
              ↑
              └──── The digit in the thousandths place (2) is
                    less than 5, so we eliminate it.
   ```

   0.0925 (rounded to the hundredths place) = 0.09

Round off these numbers to the indicated places.

a.   Round to the tenths place:

21.246 __*21.2*__

b.   Round to the tenths place:

0.0986 __*0.01*__

c.   Round to the hundredths place:

482.0146 __*482.01*__

d.   Round to the hundredths place:

0.0896 __*0.09*__

---

2. Round off these numbers to the indicated places.

a.   Round to the tenths place:

42.863 __*42.9*__

b.   Round to the hundredths place:

0.0649 __*0.06*__

c.   Round to the hundredths place:

0.0101 __*0.01*__

d.   Round to the hundredths place:

0.0991 __*0.01*__

1. a. 21.2
   b. 0.1
   c. 482.01
   d. 0.09

---

3. a.   When rounding off numbers, if the digit immediately to the right of the place we wish to round off to is 5 or more, we

_____ .

b.   If the digit to the right is 4 or less, we _____ .

2. a. 42.9
   b. 0.06
   c. 0.01
   d. 0.10

---

4. The precision of any measurement is based on the smallest subdivision of the instrument used to make the measurement.

Examples: a.   The measurement 951 is precise to the ones place.
           b.   The measurement 10.1 is precise to the tenths place.
           c.   The measurement 0.013 is precise to the thousandths place.

What is the precision of these numbers?

a.   3.18 _____

b.   15.3 _____

c.   0.072 _____

3. a. add 1 to the number
      in the place being
      rounded off to.
   b. eliminate the digit.

---

5. Sometimes a number is precise to a certain place, but the number in that position is zero. The answer *must* reflect the precision of the measuring device, even if it is necessary to use a zero in one or more places.

Examples: a.   The measurement 29.00 indicates that the measuring device can measure to the hundredths place, even though the measurement was zero in the hundredths place.
           b.   The measurement 40.100 indicates that the measuring device can measure to the thousandths place.

To what place are these numbers precise?

a.   3.10 _____

b.   0.0700 _____

c.   42.010 _____

4. a. hundredths place
   b. tenths place
   c. thousandths place

6. In adding or subtracting measurements, an answer cannot be more precise than the least precise measurement used to arrive at the answer.

Examples: a.   28.14
　　　　　　+ <u>10.2</u>   ← least precise measurement
　　　　　　38.34   ← answer must be rounded off to the tenths place
　　　　　　38.34 (rounded off) = 38.3

　　　　　　b.   1.00   ← least precise measurement
　　　　　　+ <u>0.0093</u>
　　　　　　1.0093   ← answer must be rounded off to the hundredths place
　　　　　　1.0093 (rounded off) = 1.01

　　　　　　c.   14.62
　　　　　　− <u>3.0</u>   ← least precise measurement
　　　　　　11.62   ← answer must be rounded off to the tenths place
　　　　　　11.62 (rounded off) = 11.6

Solve these problems.

a.　104.31　　b.　58.010　　c.　34.0　　d.　27.6512
　+ <u>22.111</u>　　+ <u>12.00</u>　　− <u>12.15</u>　　− <u>12.34</u>

Part 3. Determining Significant Digits

1. Not all digits in a number are significant; some digits are merely placeholders.

Example: In the number 0.0097, all digits to the left of the 9 are placeholders and therefore not significant.

There are rules for determining what digits are significant in a number. *All nonzero digits are significant.*

Examples: a.   897 has three significant digits.
　　　　　　b.   58.62 has four significant digits.

Do not let the decimal point in 58.62 disturb you. When you are counting significant digits, you can ignore the decimal point.

How many significant digits are in these numbers?

a.   29 _____　　b.   3.14 _____　　c.   123987 _____

---

5. a.   hundredths place
   b.   ten-thousandths place
   c.   thousandths place

6. a.   126.42
   b.   70.01
   c.   21.9
   d.   15.31

2. *Zeros at the left of a measurement are not significant.*

   Examples: a.   0.0097 has only two significant digits; the 9 and the 7.
               b.   0.0193 has three significant digits.

   How many significant digits are in these numbers?

   a.   0.001 _____   b.   0.0987 _____   c.   0.000034 _____

---

3. How many significant digits are in these numbers?

   a.   657 _____   b.   0.0657 _____

   c.   0.009 _____   d.   4.26 _____

---

4. *Zeros between nonzero digits are significant.*

   Examples: a.   903 has three significant digits.
               b.   1.04 has three significant digits.
               c.   0.0204 has three significant digits.

   How many significant digits are in these numbers?

   a.   406 _____   b.   28.07 _____

   c.   0.00109 _____   d.   1.002 _____

---

5. How many significant digits are in these numbers?

   a.   509 _____   b.   4.29 _____

   c.   0.06 _____   d.   0.101 _____

---

6. *Zeros at the right end of a number, after a decimal point, are significant.*

   Examples: a.   0.400 has three significant digits.
               b.   0.0090 has two significant digits.
               c.   5.000 has four significant digits.

   How many significant digits are there in these numbers?

   a.   0.5000 _____   b.   0.0080 _____

   c.   6.090 _____   d.   7.00 _____

---

7. How many significant digits are in the following numbers?

   a.   0.0082 _____   b.   907 _____   c.   0.0104 _____

   d.   2.01 _____   e.   0.30 _____   f.   0.010 _____

   g.   6.090 _____   h.   914 _____   i.   6.00 _____

---

8. *Zeros to the right of a nonzero digit but before the decimal point may or may not be significant.*

   Examples: a.   400 (The zeros may be significant.)
               b.   6000 (The zeros may be significant.)

   In this book you may assume that these digits *are* significant. In other circumstances you will have to be told whether they are significant or not.

Answers column:

1. a. two
   b. three
   c. six

2. a. one
   b. three
   c. two

3. a. three
   b. three
   c. one
   d. three

4. a. three
   b. four
   c. three
   d. four

5. a. three
   b. three
   c. one
   d. three

6. a. four
   b. two
   c. four
   d. three

7. a. two
   b. three
   c. three
   d. three
   e. two
   f. two
   g. four
   h. three
   i. three

9. In multiplying and dividing measurements, an answer cannot have more significant digits than the measurement with the least number of significant digits used to arrive at the answer.

Examples: a. $48.227 \times 1.64 = 79.1$

measurement with ⎯⎯⎦   ⎣⎯ answer must have the same
the least number   number of digits as the measure-
of significant digits.   ment with the least number of
                                          significant digits.

The answer has been rounded off to three digits.

   b. $\dfrac{52.7}{0.60} = 88$

answer must have the same number of digits as the
measurment with the least number of significant digits
measurement with the least number of significant digits

The answer has been rounded off to two digits.

Here are some problems with answers. Round the answers to the proper number of significant digits.

a. $5.237 \times 6.4 = 33.5168$ _____   b. $2.309 \times 35.6 = 82.2004$ _____

c. $\dfrac{58.7}{2.4} = 24.45833$ _____   d. $\dfrac{0.5178}{0.087} = 5.952$ _____

---

10. What happens if you want to use numbers that are not measurements along with numbers that are? Numbers that are not measurements are called *exact numbers.* Exact numbers do not have significant digits and are ignored when determining the number of significant digits in an answer.

9. a. 34
   b. 82.2
   c. 24
   d. 6.0

Examples: a. Divide the measurement 44.88 by 2.

four significant digits → $\dfrac{44.88}{2} = 22.44$
no significant digits   →

⎣⎯ The answer has four significant digits.

   b.   Multiply the measurement 10.2 by 4.

10.2 ← three significant digits
$\times \quad 4$ ← no significant digits

40.8 ← The answer has three significant digits.

Perform the indicated operation.

a.   Multiply the measurement 50.6 by 4.

b.   Divide the measurement 8.842 by 2.

11. Here is a summary of the rules for using the concept of significant digits.

Precision

In adding or subtracting measurements, an answer cannot be more precise than the least precise measurement used to arrive at the answer. We are concerned only with the number of decimal places in the answer.

Significant Digits

In multiplying and dividing measurements, an answer cannot have more significant digits than the measurement with the least number of significant digits used to arrive at the answer. We are interested only in the number of significant digits in the answer.

12. Solve these problems.

To what place are these numbers precise?

a.   14.04 _____   b.   309.1 _____

c.   0.006 _____   d.   29.00 _____

Round off these numbers to the indicated places.

e.   6.429 to the tenths place = _____

f.   5.009 to the hundredths place = _____

Add or subtract these measurements and give the answer to the correct decimal place.

g.   2.0   h.   6.944   i.   34.000
     + 3.88   − 3.33   − 2.15

How many significant digits are in the following numbers?

j.   0.0067 _____   k.   409 _____   l.   67.02 _____   m. 0.0104 _____

n.   938 _____   o.   43.00 _____   p.   0.3 _____   q.   0.0120 _____

Round off these answers to the correct number of significant digits.

r.   5.386 × 2.11 = 11.36446 _____

s.   4.090 × 28.1 = 114.9290 _____

t. $\frac{68.00}{24.0}$ = 2.83333 _____

u. $\frac{0.6140}{0.093}$ = 6.60215 _____

v.   Multiply the measurement 21.3 × 3 (3 is exact).

     _____

w.   Divide the measurement 6.864 by 2 (2 is exact).

13. Answers to frame 12.

a. hundredths	e. 6.4	i. 31.85	m. three	q. three	u. 6.6
b. tenths	f. 5.01	j. two	n. three	r. 11.4	v. 63.9
c. thousandths	g. 5.9	k. three	o. four	s. 115	w. 3.432
d. hundredths	h. 3.61	l. four	p. one	t. 2.83	

10. a. 202
    b. 4.421

Part 4. Dimensional Analysis

1. Both numbers and dimensions must be evaluated to solve problems in chemistry. We have learned how to evaluate numbers, and now we will learn how to evaluate dimensions. The process of evaluating dimensions in formulas is called dimensional analysis.

   The concept most often used in dimensional analysis is:

   $$\frac{n}{n} = 1$$

   This concept is used in a short-cut technique called canceling. Both numbers and dimensions can be canceled.

   Example: $\dfrac{3 P V}{3 P} = V$

   The 3s and *P*s canceled because they appeared in both the numerator and denominator and they take the form:

   $$\frac{3}{3} = 1 \quad \text{and} \quad \frac{P}{P} = 1$$

   Here are other examples of reducing by canceling.

   a. $\dfrac{\overset{2}{10} L K}{5 K} = 2 L$ b. $\dfrac{(\overset{2}{6} K)(4 L)}{3 K} = 8 L$

   Note that when numbers are canceled they do not always equal one but are reduced to lowest terms.

   Dimensional analysis consists of canceling units of dimensions in a formula to give an answer with the correct dimensional units.

   Use dimensional analysis to reduce the following equations.

   a. $\dfrac{5 \text{ m cm}}{5 \text{ m}} =$ b. $\dfrac{1000 \text{ mg g}}{500 \text{ mg}} =$

   c. $\dfrac{500 \text{ cm m}}{1000 \text{ cm}} =$ d. $\dfrac{(2 L)(4 K)}{(2 K)} =$

   e. $\dfrac{1000 \text{ mL } L}{100 \text{ mL}} =$ f. $\dfrac{(10 L)(6 K)}{5 K} =$

2. Dimensional analysis can also be used to evaluate formulas written in conversion-factor form.

Example: Reduce the following.

$$\cancel{100} \; \cancel{cm} \times \frac{1 \text{ m}}{\cancel{100} \; \cancel{cm}} = 1 \text{ m}$$

This can be done because

$$100 \text{ cm} \times \frac{1 \text{ m}}{100 \text{ cm}} = \frac{(100 \text{ cm}) \; (1 \text{ m})}{100 \text{ cm}}$$

We do not bother to rewrite the conversion-factor form.

Examples: a. $2 \; \cancel{cm} \times \dfrac{10 \text{ mm}}{1 \; \cancel{cm}} = 20 \text{ mm}$

b. $\overset{5}{\cancel{500}} \; \cancel{cm} \times \dfrac{1 \text{ m}}{\cancel{100} \; \cancel{cm}} = 5 \text{ m}$

Use dimensional analysis to reduce the following conversion-factor equations.

a. $80 \text{ mm} \times \dfrac{1 \text{ cm}}{10 \text{ mm}} =$

b. $2000 \text{ mL} \times \dfrac{1 \text{ L}}{1000 \text{ mL}} =$

c. $2 \text{ m} \times \dfrac{100 \text{ cm}}{1 \text{ m}} =$

d. $20 \text{ cm} \times \dfrac{10 \text{ mm}}{1 \text{ cm}} =$

e. $500 \text{ g} \times \dfrac{1 \text{ kg}}{1000 \text{ g}} =$

f. $200 \text{ mL} \times \dfrac{1 \text{ L}}{1000 \text{ mL}} =$

3. Here are more-complex conversion-factor equations. They are reduced in the same manner using dimensional analysis for the units and canceling numbers.

Example: $200 \; \cancel{mL} \times \dfrac{1 \; \cancel{kg}}{100 \; \cancel{mL}} \times \dfrac{1000 \text{ g}}{1 \; \cancel{kg}} = 2000 \text{ g}$

Try these.

a. $600 \; K \times \dfrac{1 \, P}{300 \, K} \times \dfrac{10 \, V}{4 \, P} =$

b. $100 \; L \times \dfrac{10 \, V}{2 \, L} \times \dfrac{10 \, K}{5 \, V} =$

4. Answers to frame 3.

a. $5 \, V$ b. $1000 \, K$

1. a. cm
   b. 2 g
   c. 0.5 m
   d. 4 *L*
   e. 10 *L*
   f. 12 *L*

2. a. 8 cm
   b. 2 *L*
   c. 200 cm
   d. 200 mm
   e. 0.5 kg
   f. 0.2 *L*

NAME _____

MEASUREMENT CONCEPTS AND
DIMENSIONAL ANALYSIS
EVALUATION TEST
PART 1–PART 4

Round off the answers of the added or subtracted quantities to show the correct precision.

1. $0.0916 + 142.8 + 9.56 = 152.4516$

2. $1.4261 - 0.506 + 0.81537 = 1.73547$

3. $12.0 + 7.850 - 4.20 = 15.650$

State the number of significant digits in each measurement.

4. 0.0078 5. 0.06080 6. 4.09090

Round off the answers of the multiplied or divided quantities to the correct number of significant digits.

7. $19.74 \times 0.863 \times 46.35 = 789.6009870$

8. $\dfrac{79.0}{37.06} = 2.131678$

9. $\dfrac{617.1 \times 0.532}{326.94} = 1.004151$

10. $\dfrac{328 \times 0.0083}{61.92 \times 0.0144} = 3.05322$

Use dimensional analysis to reduce the following conversion-factor equations.

11. $\dfrac{15 \text{ m cm}}{5 \text{ m}} =$

12. $\dfrac{(20 \text{ L})(6 \text{ K})}{2 \text{ K}} =$

13. $40 \text{ mm} \times \dfrac{1 \text{ cm}}{10 \text{ mm}} =$

14. $100 \text{ mL} \times \dfrac{1 \text{ L}}{1000 \text{ mL}} =$

15. $400 \text{ mL} \times \dfrac{1 \text{ kg}}{100 \text{ mL}} \times \dfrac{1000 \text{ g}}{1 \text{ kg}} =$

1. _____
2. _____
3. _____
4. _____
5. _____
6. _____
7. _____
8. _____
9. _____
10. _____
11. _____
12. _____
13. _____
14. _____
15. _____

UNIT 5
The Metric System, Density and Specific Gravity

In this unit you will learn to use the metric system of measurement. You will learn the basic organization of the system and how to measure length, volume, and mass using the metric system. You will also learn the structure of the Fahrenheit and Celsius temperature scales and how to convert from one temperature scale to the other. You will learn to calculate the density and specific gravity of a substance.

Part 1. Introduction

The metric system or International System (SI) is used by almost all countries in the world. The International System measures many different quantities as shown in the table. We will only use a few of these units in this book and will refer to the International System as the metric system.

International System Base Units of Measurement

Quantity	Name of Unit
Length	Meter
Mass	Kilogram
Temperature	Kelvin
Time	Second
Amount of substance	Mole
Electric current	Ampere
Luminous intensity	Candela

Part 2. Basic Organization of the Metric System

1. A good measuring system (a) has easily reproducible standards, (b) is based on a convenient number, and (c) has an interrelationship between the different units. The metric system of measurement was designed to meet all of these criteria.

2. The metric system is an efficient system of measurement because it is based on the number 10. This means that all multiples and subdivisions of the standard unit also are based on the number 10. The English system of measurement, which has always been used in the United States, is not based on any one number.

   Example: standard unit of length: 1 foot

   1 foot = 12 inches (subdivision)
   1 yard = 3 feet (multiple)

Note that subdivisions of the standard foot are based on the number 12 and multiples of the foot are based on the number 3. If the English system of measurement were based on the number 10, as the metric system is, all subdivisions and multiples would also be based on the number 10.

Example: standard unit of length: 1 foot
                1 foot = 10 inches (subdivision)
                1 yard = 10 feet (multiple)

A measuring system based entirely on the number 10 is both consistent and easy to work with in solving problems.

The metric system is based on the number _____ .

---

3. The English system is based on _____ .

2. 10

---

4. If the English system were based on the number 10, there would be a multiple of _____ ounces in a quart.

3. many numbers

---

5. If the English system were based on the number 10, there would be a multiple of _____ quarts in a gallon.

4. 10

---

6. A prefix before the name of a unit in the metric system is used to indicate what multiple or subdivision is wanted. These prefixes are listed in the following table.

5. 10

Prefix	Meaning		Prefix	Meaning	
micro	1000000 of these in a standard unit		deka	10 times the standard unit	
milli	1000 of these in a standard unit	subdivisions	hecto	100 times the standard unit	multiples
centi	100 of these in a standard unit		kilo	1000 times the standard unit	
deci	10 of these in a standard unit				

You must know these prefixes.

Tell how many of these subdivisions are in a standard unit.

a.   deci _____   b.   milli _____   c.   centi _____

---

7. Tell how many of the standard units there are in each multiple.

a.   kilo _____   b.   deka _____   c.   hecto _____

6. a. 10
   b. 1000
   c. 100

---

8. a.   If there are 1000 subdivisions in the standard unit, the prefix is _____ .

   b.   For 1000 standard units in a multiple, the prefix is _____ .

   c.   If there are 100 subdivisions in the standard unit, the prefix is _____ .

7. a. 1000
   b. 10
   c. 100

---

9. Tell whether the prefix represents a multiple or a subdivision of the standard unit.

a.   deci _____

b.   kilo _____

c.   milli _____

8. a. milli
   b. kilo
   c. centi

Part 3. The Measurement of Length

1. The standard unit of length in the metric system is the meter. The meter is a unit of length slightly larger than a yard, or 39.4 inches.

   Approximately how many feet are in a meter? _____

2. The meter is useful for measuring objects that have a similar length, but it is too long to measure very short objects and too short to measure long distances. To make a shorter measuring unit, we divide the meter into subdivisions based on the number 10. To make a longer measuring unit, we multiply the meter by 10. The table below contains the most important subdivisions and multiples of the standard meter. The dekameter and hectometer are missing, because they are not convenient units of measurement.

   You must learn the name and abbreviation of each of the following subdivisions and multiples and its relationship to the standard unit. Notice that the abbreviations are not capitalized and do not use periods.

Unit	Abbreviation	Subdivisions and Multiples
millimeter	mm	1000 of these in a meter
centimeter	cm	100 of these in a meter
decimeter	dm	10 of these in a meter
meter	m	standard unit
kilometer	km	1000 meters

   39.4 in

   Give the abbreviations of the following units.

   a. millimeter _____   b. meter _____   c. kilometer _____

   d. decimeter _____   e. centimeter _____

3. Give the name for each of these abbreviations.

   a. cm _____   b. dm _____

   c. m _____   d. km _____

   e. mm _____

4. How many of the following are contained in a given unit?

   a. In one meter there are _____ mm.

   b. In one meter there are _____ dm.

   c. In one meter there are _____ cm.

   d. In one kilometer there are _____ m.

5. It is easy to remember how many centimeters there are in one meter. It becomes more difficult to know how many centimeters there are in 2.85 meters. To solve this problem, we use conversion factors. To use conversion factors, we must know how to cancel units as well as numbers. In canceling numbers, we rely on the fact that a number divided by itself equals 1.

   Examples: $\frac{3}{3} = 1$ $\frac{79}{79} = 1$

----

9. a. subdivision
    b. multiple
    c. subdivision

1. 3

2. a. mm
    b. m
    c. km
    d. dm
    e. cm

3. a. centimeter
    b. decimeter
    c. meter
    d. kilometer
    e. millimeter

4. a. 1000
    b. 10
    c. 100
    d. 1000

The same holds true for units.

Examples: $\dfrac{m}{m} = 1$ $\dfrac{cm}{cm} = 1$

If we have a more complex situation, we rearrange the numbers.

Example: $12 \times \dfrac{8}{12} = 8 \times \dfrac{12}{12} = 8 \times 1 = 8$

We can do this same operation with units.

$cm \times \dfrac{mm}{cm} = mm \times \dfrac{cm}{cm} = mm \times 1 = mm$

$m \times \dfrac{km}{m} = km \times \dfrac{m}{m} = km \times 1 = km$

Try these problems.

a. $m \times \dfrac{dm}{m} =$

b. $mm \times \dfrac{cm}{mm} =$

c. $cm \times \dfrac{m}{cm} =$

---

6. We can also cancel units in a more abbreviated manner. You need not rearrange the problem. Simply cancel the units that are alike.

Example: $2.1 \ cm \times \dfrac{10 \ mm}{1 \ cm}$

Step 1. Cancel like units.

$2.1 \ \cancel{cm} \times \dfrac{10 \ mm}{1 \ \cancel{cm}}$

Step 2. Multiply the numbers and add the remaining unit to the answer.

$2.1 \times 10 \ mm = 21 \ mm$

Example: $321 \ cm \times \dfrac{1 \ m}{100 \ cm}$

Step 1. Cancel like units.

$321 \ \cancel{cm} \times \dfrac{1 \ m}{100 \ \cancel{cm}}$

Step 2. Divide the numbers and add the remaining unit to the answer.

$\dfrac{321}{100} \times 1 \ m = 3.21 \ m$

Note that in the first example we multiplied the numbers and in the second example we divided the numbers. When the larger numerical value is at the top in the fraction, we multiply; when it is at the bottom, we divide.

Try these problems.

a. $4.618 \ km \times \dfrac{1000 \ m}{1 \ km} =$ b. $32.5 \ m \times \dfrac{100 \ cm}{1 \ m} =$

c. $1400 \ m \times \dfrac{1 \ km}{1000 \ m} =$ d. $1860 \ mm \times \dfrac{1 \ m}{1000 \ mm} =$

5. a. $dm \times \dfrac{m}{m} = dm \times 1 = dm$

b. $cm \times \dfrac{mm}{mm} = cm \times 1 = cm$

c. $m \times \dfrac{cm}{cm} = m \times 1 = m$

7. Look carefully at the way the conversion factor is written. Note that the unit we want to convert *to* is at the top of the conversion factor and the unit we want to convert *from* is at the bottom of the conversion factor. If the conversion factor is not written correctly, the unit you want to get rid of will not take the proper form—for example, $\frac{m}{m} = 1$.

Example:

$$2.85 \text{ m} \times \frac{100 \text{ cm}}{1 \text{ m}}$$

— unit desired at top

— unit to be changed at bottom

Which of the following are set up correctly? _____

a. $2.6 \text{ km} \times \dfrac{1000 \text{ m}}{1 \text{ km}}$

b. $10 \text{ cm} \times \dfrac{10 \text{ mm}}{1 \text{ cm}}$

c. $182 \text{ cm} \times \dfrac{100 \text{ cm}}{1 \text{ m}}$

d. $82 \text{ mm} \times \dfrac{1000 \text{ mm}}{1 \text{ m}}$

8. Let's go through a conversion problem step by step.

Example: Change 2.85 m to centimeters.

Step 1. Set up the conversion-factor formula.

$$2.85 \text{ m} \times \frac{\text{cm}}{\text{m}}$$

Step 2. Now determine how many of the smaller units are in the larger unit.

smaller unit = cm
larger unit = m

There are 100 cm in one meter. Put this information into the conversion-factor formula.

$$2.85 \text{ m} \times \frac{100 \text{ cm}}{1 \text{ m}}$$

Step 3. Solve the problem.

$$2.85 \text{ m} \times \frac{100 \text{ cm}}{1 \text{ m}} = 285 \text{ cm}$$

Example: Change 4870 mm to meters.

Step 1. Set up the formula.

$$4870 \text{ mm} \times \frac{\text{m}}{\text{mm}}$$

Step 2. smaller unit = mm
larger unit = m

There are 1000 mm in one meter. Insert this information into the formula.

$$4870 \text{ mm} \times \frac{1 \text{ m}}{1000 \text{ mm}}$$

Step 3. Solve the problem.

$$4870 \text{ mm} \times \frac{1 \text{ m}}{1000 \text{ mm}} = 4.870 \text{ m}$$

6. a. 4618 m
   b. 3250 cm
   c. 1.400 km
   d. 1.860 m

7. a and b

Try this problem, using the three steps. Change 8690 m to kilometers.

Step 1.

Step 2.

Step 3.

---

9. Change 0.03 m to millimeters.

   Step 1.

   Step 2.

   Step 3.

---

10. We must often change measurements from centimeters to millimeters.

    Example: Change 28 mm to centimeters (1 cm = 10 mm).

    Step 1. 28 mm $\times$ $\dfrac{cm}{mm}$

    Step 2. smaller unit = mm
             larger unit = cm
             1 cm = 10 mm

             28 mm $\times$ $\dfrac{1 \text{ cm}}{10 \text{ mm}}$

    Step 3. 28 mm $\times$ $\dfrac{1 \text{ cm}}{10 \text{ mm}}$ = 2.8 cm

    Change 4.2 cm to millimeters.

    Step 1.

    Step 2.

    Step 3.

---

11. Set up but *do not solve* the following conversion equations.

    a. Change 10 m to decimeters: 10 m $\times$ ———

    b. Change 100 m to centimeters: 100 m $\times$ ———

    c. Change 1000 mm to meters: 1000 mm $\times$ ———

---

8. Step 1.
   8690 m $\times$ $\dfrac{km}{m}$

   Step 2.
   8690 m $\times$ $\dfrac{1 \text{ km}}{1000 \text{ m}}$

   Step 3.
   8690 m $\times$ $\dfrac{1 \text{ km}}{1000 \text{ m}}$ = 8.690 km

9. Step 1.
   0.03 m $\times$ $\dfrac{mm}{m}$

   Step 2.
   0.03 m $\times$ $\dfrac{1000 \text{ mm}}{1 \text{ m}}$

   Step 3.
   0.03 m $\times$ $\dfrac{1000 \text{ mm}}{1 \text{ m}}$ = 30 mm

10. Step 1.
    4.2 cm $\times$ $\dfrac{mm}{cm}$

    Step 2.
    4.2 cm $\times$ $\dfrac{10 \text{ mm}}{1 \text{ cm}}$

    Step 3.
    4.2 cm $\times$ $\dfrac{10 \text{ mm}}{1 \text{ cm}}$ = 42 mm

12. Here are more problems for you to set up.

    a. Change 0.5 m to centimeters.

    b. Change 2000 mm to meters.

    c. Change 5 cm to millimeters.

    d. Change 860 m to kilometers.

---

13. Solve these problems:

    a. Change 0.2 m to centimeters.

    b. Change 1.516 m to millimeters.

    c. Change 150 cm to meters.

    d. Change 15 mm to meters.

    e. Change 138 mm to centimeters.

    f. Change 0.89 cm to millimeters.

    g. Change 2.312 km to meters.

11. a. $10 \text{ m} \times \dfrac{10 \text{ dm}}{1 \text{ m}}$

    b. $100 \text{ m} \times \dfrac{100 \text{ cm}}{1 \text{ m}}$

    c. $1000 \text{ mm} \times \dfrac{1 \text{ m}}{1000 \text{ mm}}$

12. a. $0.5 \text{ m} \times \dfrac{100 \text{ cm}}{1 \text{ m}}$

    b. $2000 \text{ mm} \times \dfrac{1 \text{ m}}{1000 \text{ mm}}$

    c. $5 \text{ cm} \times \dfrac{10 \text{ mm}}{1 \text{ cm}}$

    d. $860 \text{ m} \times \dfrac{1 \text{ km}}{1000 \text{ m}}$

Part 4. The Measurement of Volume

1. Another type of measurement is that of volume. Volume measurements are derived from linear measurements by cubing the linear measurements. Thus, the volume of a box, for example, is found by multiplying the length times the width times the height of the box.

    volume of box = $l \times w \times h$

    If we use a box that is 10 cm on each edge, we have the following:

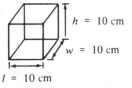

h = 10 cm
w = 10 cm
l = 10 cm

volume = 10 cm × 10 cm × 10 cm
volume = 10 × 10 × 10 = 1000
      cm × cm × cm = cm³ } = 1000 cm³

(This number is read as 1000 cubic centimeters and is often abbreviated as 1000 cc.)

As you can see, we cube not only the number but also the units.

Because the unit of volume is derived from the unit of length in the metric system, there is an interrelationship between the two. There is no such relationship in the English system of measurement.

Volume measurement is really _____ measurement cubed.

13. a. 20 cm
    b. 1516 mm
    c. 1.5 m
    d. 0.015 m
    e. 13.8 cm
    f. 8.9 mm
    g. 2312 m

---

2. In converting linear measurements to volume measurements, we cube not only the numbers but also the _____ .

1. linear

3. The volume contained in 1000 cm³ is the standard unit of volume in the metric system. The name given to this standard unit is the liter. The liter is similar to the quart in the English system. Below is a table of the two most frequently used units of volume in the metric system; the other units are not convenient to use.

Unit	Abbreviation	Subdivision
liter	L	standard unit
milliliter	mL	1000 of these in a liter

Give the name or abbreviation of each.

a. L _____   b. mL _____

c. liter _____   d. milliliter _____

4. In one liter there are _____ mL.

5. The linear measurements 10 cm × 10 cm × 10 cm equal 1000 cm³, which equals 1 L. A liter also contains 1000 mL. We can relate cm³ and mL this way.

$$1000 \text{ cm}^3 = 1 \text{ L} = 1000 \text{ mL}$$
$$\text{or} \quad 1000 \text{ cm}^3 = 1000 \text{ mL}$$
$$\text{or} \quad 1 \text{ cm}^3 = 1 \text{ mL}$$

(Remember that the equation can also be written as 1 cc = 1mL.)

1 mL = _____ cm³

6. If 1 cm³ equals 1 mL, then 10 cm³ must equal _____ mL.

7. A cm³ is often called a _____ (abbreviation).

8. We can change from one unit of volume to another by using the conversion-factor method.

Example: Change 4211 mL to liters.

Step 1. 4211 mL × $\dfrac{L}{mL}$

Step 2. smaller unit = mL   There are 1000 mL per liter.
         larger unit = L

         4211 mL × $\dfrac{1 \ L}{1000 \text{ mL}}$

Step 3. 4211 mL × $\dfrac{1 \ L}{1000 \text{ mL}}$ = 4.211 L

Try these problems.

a. Change 8976 mL to liters.   b. Change 2.600 L to milliliters.

c. Change 88 mL to cubic centimeters.   d. Change 1400 mL to liters.

---

2. units

3. a. liter
   b. milliliter
   c. L
   d. mL

4. 1000

5. 1

6. 10

7. cc

Part 5. The Measurement of Mass

1. The remaining quantity that we want to measure is that of mass, which is often confused with weight. Mass is the quantity of matter an object possesses. Weight is caused by the pull of gravity on the mass of an object. The mass of an object is constant, but the weight of an object depends on the distance an object is from the center of the earth. If an object is far enough from the center of the earth, it becomes weightless, but it still retains its mass. Look at the illustration.

8. a. 8976 mL × $\dfrac{1\ L}{1000\ mL}$
       = 8.976 L

    b. 2.600 L × $\dfrac{1000\ mL}{1\ L}$
       = 2600 mL

    c. 88 mL × $\dfrac{1\ cm^3}{1\ mL}$
       = 88 cm^3

    d. 1400 mL × $\dfrac{1\ L}{1000\ mL}$
       1.400 L

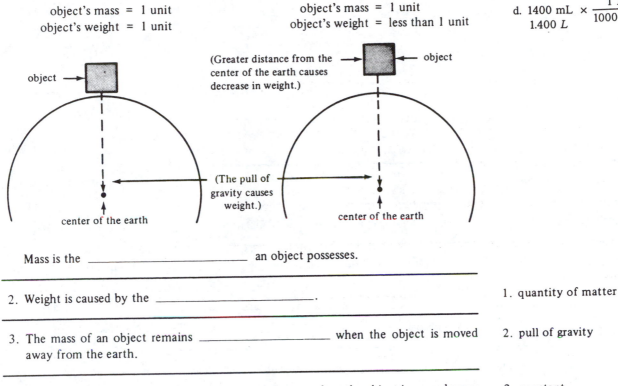

object's mass = 1 unit
object's weight = 1 unit

object's mass = 1 unit
object's weight = less than 1 unit

(Greater distance from the center of the earth causes decrease in weight.)

object

object

(The pull of gravity causes weight.)

center of the earth

center of the earth

Mass is the _____ an object possesses.

2. Weight is caused by the _____.

3. The mass of an object remains _____ when the object is moved away from the earth.

4. The weight of an object _____ when the object is moved away from the earth.

5. Mass and weight are measured in different ways. We measure weight by putting a scale between the object and the center of the earth and letting the gravity pull the object down on the scale.

1. quantity of matter

2. pull of gravity

3. constant

4. decreases

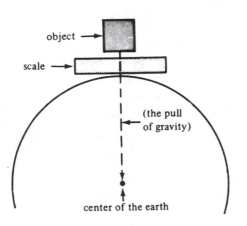

object

scale

(the pull of gravity)

center of the earth

We do not use scales in a chemistry laboratory, because chemistry is concerned with mass rather than weight. We measure mass by balancing one object against a standard of mass. The effect of gravity is canceled because it acts equally on each side of the balance, as in the following illustration.

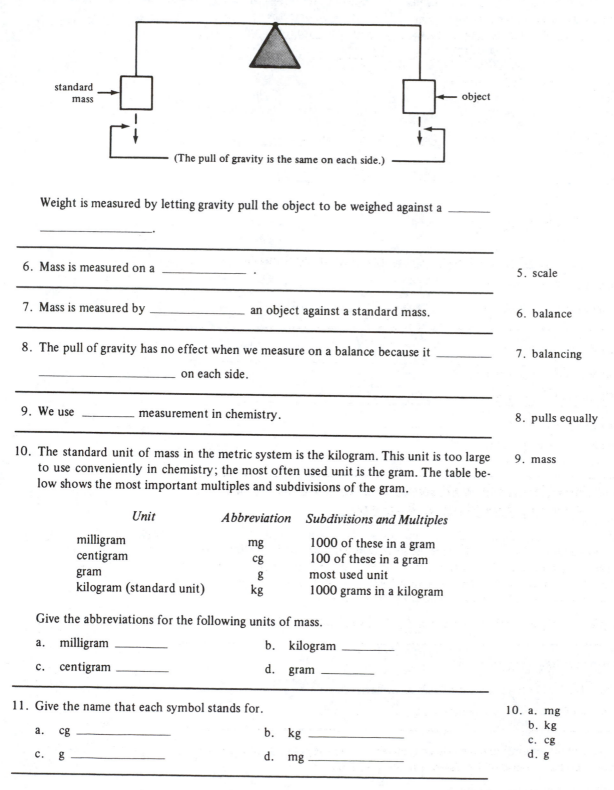

(The pull of gravity is the same on each side.)

Weight is measured by letting gravity pull the object to be weighed against a _____ _____.

---

6. Mass is measured on a _____ .

5. scale

---

7. Mass is measured by _____ an object against a standard mass.

6. balance

---

8. The pull of gravity has no effect when we measure on a balance because it _____ _____ on each side.

7. balancing

---

9. We use _____ measurement in chemistry.

8. pulls equally

---

10. The standard unit of mass in the metric system is the kilogram. This unit is too large to use conveniently in chemistry; the most often used unit is the gram. The table below shows the most important multiples and subdivisions of the gram.

9. mass

Unit	Abbreviation	Subdivisions and Multiples
milligram	mg	1000 of these in a gram
centigram	cg	100 of these in a gram
gram	g	most used unit
kilogram (standard unit)	kg	1000 grams in a kilogram

Give the abbreviations for the following units of mass.

a. milligram _____   b. kilogram _____

c. centigram _____   d. gram _____

---

11. Give the name that each symbol stands for.

a. cg _____   b. kg _____

c. g _____   d. mg _____

10. a. mg
    b. kg
    c. cg
    d. g

12. How many of the following are contained in the given unit?

    a. In one gram there are _____ cg.

    b. In one gram there are _____ mg.

    c. In one kilogram there are _____ g.

11. a. centigram
    b. kilogram
    c. gram
    d. milligram

13. Use the conversion-factor method to convert from one unit of mass to another.

    Example: Change 1285 mg to grams.

    Step 1. $1285 \text{ mg} \times \dfrac{g}{mg}$

    Step 2. smaller unit = mg   There are 1000 mg per gram.
              larger unit = g

    $$1285 \text{ mg} \times \frac{1 \text{ g}}{1000 \text{ mg}}$$

    Step 3. $1285 \text{ mg} \times \dfrac{1 \text{ g}}{1000 \text{ mg}} = 1.285 \text{ g}$

    Set up, but do not solve, the conversion-factor equations for the following.

    a. Change 100 g to kilograms.

    b. Change 10 g to milligrams.

12. a. 100
    b. 1000
    c. 1000

14. Solve these problems.

    a. Change 2.215 kg to grams.

    b. Change 489 g to kilograms.

    c. Change 4.265 g to milligrams.

    d. Change 920 mg to grams.

13. a. $100 \text{ g} \times \dfrac{1 \text{ kg}}{1000 \text{ g}}$

    b. $10 \text{ g} \times \dfrac{1000 \text{ mg}}{1 \text{ g}}$

Part 6. Relationship Between the Mass and Volume of Water in the Metric System

1. If you take a balance and find the mass of 1 mL of water, you will find that it measures approximately one gram. For most practical work we can use the measurement: 1 mL of water has a mass of 1 g. This is a very handy concept because if you know the volume of water, you also know its mass, and vice versa.

    Example: 1 mL of water has a mass of 1 g, so 10 mL of water have a mass of 10 g.

4. a. $2.215 \text{ kg} \times \dfrac{1000 \text{ g}}{1 \text{ kg}}$
       $= 2215 \text{ g}$

   b. $489 \text{ g} \times \dfrac{1 \text{ kg}}{1000 \text{ g}}$
       $= 0.489 \text{ kg}$

   c. $4.265 \text{ g} \times \dfrac{1000 \text{ mg}}{1 \text{ g}}$
       $= 4265 \text{ mg}$

   d. $920 \text{ mg} \times \dfrac{1 \text{ g}}{1000 \text{ mg}}$
       $= 0.920 \text{ g}$

2. Complete the following statements.

    a. 28 mL of water have a mass of _____ g.

    b. 19.5 g of water have a volume of _____ mL.

2. a. 28
   b. 19.5

3. Because a milliliter (mL) and a cubic centimeter (cc) are the same, a cubic centimeter of water also has a mass of 1 g. Answer these questions.

    a. 28.2 cc of water have a mass of _____ g.

    b. 200 g of water have a volume of _____ cc.

Part 7. Conversion Between the English and Metric Systems

1. At times it is necessary to convert from the English to the metric system and vice versa. This is a list of some of the most commonly used units in each system and the relationship between the units.

Length			Volume		
English		*Metric*	*English*		*Metric*
1 in	=	2.54 cm	1 qt	=	0.95 *L*
1 mi	=	1.6 km	1 oz (liquid)	=	30 mL
39.4 in	=	1 m			

Mass or Weight		
English		*Metric*
1 lb	=	454 g
2.2 lb	=	1 kg
1 oz (weight)	=	28 g

Answer the following.

a. 1 in = _____ cm   b. 1 mi = _____ km   c. 1 m = _____ in

2. What is the equivalent of the following?

a. 1 qt = _____ *L*   b. 1 oz (liquid) = _____ mL

3. Answer these questions.

a. 1 mi = _____ km   b. 1 qt = _____ *L*

c. 1 in = _____ cm   d. 1 oz (liquid) = _____ mL

4. Fill in the blanks.

a. 1 lb = _____ g   b. 1 kg = _____ lb

c. 1 oz (weight) = _____ g

5. Complete these equations.

a. 1 qt = _____ *L*   b. 1 m = _____ in   c. 1 lb = _____ g

d. 1 in = _____ cm   e. 1 kg = _____ lb   f. 1 mi = _____ km

g. 1 oz (liquid) = _____ mL   h. 1 oz (weight) = _____ g

6. You can use the conversion-factor method you learned earlier to convert amounts from one system to the other.

Example: Suppose we are asked how many centimeters there are in 2 in. We know that 1 in equals 2.54 cm, so we set up a conversion equation.

─── The unit desired is on top.

$$2 \text{ in} \times \frac{2.54 \text{ cm}}{1 \text{ in}} = 5.08 \text{ cm}$$

─── The unit to be changed is on the bottom.

3. a. 28.2
   b. 200

1. a. 2.54
   b. 1.6
   c. 39.4

2. a. 0.95
   b. 30

3. a. 1.6
   b. 0.95
   c. 2.54
   d. 30

4. a. 454
   b. 2.2
   c. 28

5. a. 0.95
   b. 39.4
   c. 454
   d. 2.54
   e. 2.2
   f. 1.6
   g. 30
   h. 28

The problem is solved with the same three steps we used before.

Step 1. Set up the conversion factor.

$$2 \text{ in } \times \frac{\text{cm}}{\text{in}}$$

Step 2. Determine the relationship between the units.

$$1 \text{ in } = 2.54 \text{ cm}$$

$$2 \text{ in } \times \frac{2.54 \text{ cm}}{1 \text{ in}}$$

Step 3. Solve the problem.

$$\text{cm} \times \frac{\text{in}}{\text{in}} \times \frac{2 \times 2.54}{1} = 5 \text{ cm}$$

Set up, but do not solve, the following conversion equations.

a. Change 5 mi to kilometers.

  5 mi × ————

b. Change 5 *L* to quarts.

  5 *L* × ————

c. Change 10 lb to grams.

  10 lb × ————

d. Change 5 lb to kilograms.

  5 lb × ————

---

7. Solve these problems.

  a. Change 3.2 km to miles.

  b. Change 4.75 liters to quarts.

  c. Change 1.1 oz (liquid) to milliliters.

  d. Change 1.50 lb to grams.

  e. Change 56 g to ounces (weight).

6. a. $5 \text{ mi} \times \dfrac{1.6 \text{ km}}{1 \text{ mi}}$

  b. $5 L \times \dfrac{1 \text{ qt}}{0.95 \, L}$

  c. $10 \text{ lb} \times \dfrac{454 \text{ g}}{1 \text{ lb}}$

  d. $5 \text{ lb} \times \dfrac{1 \text{ kg}}{2.2 \text{ lb}}$

Part 8. The Fahrenheit and Celsius Temperature Systems

1. The Fahrenheit temperature scale is the system that has traditionally been used to measure temperature in the United States. The important features of this system are as follows.

7. a. 2.0 mi
   b. 5.0 qt
   c. 33 mL
   d. 681 g
   e. 2.0 oz

Fahrenheit Temperature Scale

  212° (water boils)

  32° (water freezes)

This temperature scale is divided into units called *degrees*. A degree Fahrenheit is abbreviated °F. Note that, on this scale, water freezes at 32°F and boils at 212°F. When the freezing point of water is subtracted from the boiling point of water, we find that there are 180° between the two points (212°F − 32°F = 180°F).

A unit of measure on a temperature scale is called a _____.

2. The abbreviation for *degree Fahrenheit* is _____ .

1. degree

3. 32° F is the _____ point of water.

2. °F

4. The boiling point of water is _____ °F.

3. freezing

5. There are _____ ° on the Fahrenheit scale between the freezing and boiling points of water.

4. 212

6. The Celsius temperature scale is used to measure temperature in fields of science and in most countries of the world. This system is compared with the Fahrenheit system below.

5. 180

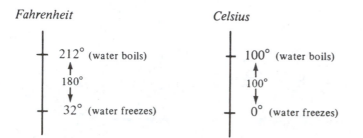

The Celsius temperature scale also is divided into units called *degrees*. A degree Celsius is abbreviated °C. On the Celsius scale, water freezes at 0° and boils at 100°. When we subtract the freezing point from the boiling point, we get 100°, which is a multiple of 10 and a much nicer number to work with than 180.

The abbreviation for *degrees Celsius* is _____ .

7. 100°C is the _____ point of water.

6. °C

8. Water freezes at _____°C.

7. boiling

9. There are _____ degrees on the Celsius scale between the freezing and boiling points of water.

8. 0

10. Looking at the diagram in frame 6, we can see that there are 100°C for every 180°F (or 1°C equals 1.8°F). We can also see that there is a 32° difference in the starting points of the scales.

9. 100

1°C = _____ °F.

11. 0°C = _____ °F.

10. 1.8

12. Which is larger—a degree Celsius or a degree Fahrenheit? _____

11. 32

_____ .

13. Formulas can be used to convert temperatures from one scale to the other. The formula for converting from Celsius to Fahrenheit is as follows.

$$°F = (1.8 \times °C) + 32$$

The formula is used in the following manner.

Example: What is 20°C on the Fahrenheit scale?

Step 1. Write the formula.

$$°F = (1.8 \times °C) + 32$$

Step 2. List the data.

$$°C = 20$$

Step 3. Substitute the data into the formula.

$$°F = (1.8 \times 20) + 32$$

Step 4. Solve the equation. (Always perform first the operation indicated within the parentheses.)

$$°F = (1.8 \times 20) + 32$$
$$°F = \quad 36 \quad + 32 = 68°F$$

Try this problem. What is 25°C on the Fahrenheit scale?

Step 1.

Step 2.

Step 3.

Step 4.

---

14. The formula for changing Fahrenheit to Celsius is as follows.

$$°C = \frac{(°F - 32)}{1.8}$$

The formula is used in the following manner.

Example: Change 50°F to °C.

Step 1. Write the formula.

$$°C = \frac{(°F - 32)}{1.8}$$

Step 2. List the data.

$$°F = 50$$

Step 3. Substitute the data into the formula.

$$°C = \frac{(50 - 32)}{1.8}$$

Step 4. Solve the equation. (Always perform first the operation indicated in parentheses.)

$$°C = \frac{(50 - 32)}{1.8} = \frac{18}{1.8} = 10°C$$

12. a degree Celsius

13. Step 1.
$$°F = (1.8 \times °C) + 32$$
Step 2.
$$°C = 25$$
Step 3.
$$°F = (1.8 \times 25) + 32$$
Step 4.
$$°F = 45 + 32 = 77°F$$

Try this problem. What is 86°F on the Celsius scale?

Step 1.

Step 2.

Step 3.

Step 4.

---

15. Sometimes negative or signed numbers are involved in converting temperatures from one scale to the other.

Example: What is −10°C on the Fahrenheit scale?

Step 1. Write the formula.

$$°F = (1.8 \times °C) + 32$$

Step 2. List the data.

$$°C = -10$$

Step 3. Substitute the data into the formula.

$$°F = (1.8 \times -10) + 32$$

Step 4. Solve the equation. (When multiplying numbers with unlike signs, the answer has a negative sign.)

$$°F = (1.8 \times -10) + 32$$
$$°F =   (-18)   + 32 = 14°F$$

Try this problem. What is −20°C on the Fahrenheit scale?

Step 1.

Step 2.

Step 3.

Step 4.

---

16. Try this problem. What is −40°F on the Celsius scale?

Step 1.

Step 2.

Step 3.

Step 4.

---

17. Solve these temperature-conversion problems.

Change the following from Celsius to Fahrenheit.

a.   30°C =   b.   −15°C =

Change the following from Fahrenheit to Celsius.

c.   32°F =   d.   23°F =

---

14. Step 1.
$$°C = \frac{(°F - 32)}{1.8}$$
Step 2.
$$°F = 86$$
Step 3.
$$°C = \frac{(86 - 32)}{1.8}$$
Step 4.
$$°C = \frac{54}{1.8} = 30°C$$

15. Step 1.
$$°F = (1.8 \times °C) + 32$$
Step 2.
$$°C = -20$$
Step 3.
$$°F = (1.8 \times -20) + 32$$
Step 4.
$$°F = (-36) + (+32) = -4°F$$

16. Step 1. $°C = \frac{(°F - 32)}{1.8}$

Step 2. $°F = -40$

Step 3. $°C = \frac{(-40) - (32)}{1.8}$

Step 4. $°C = \frac{(-40) - (+32)}{1.8}$

$$= \frac{(-40) + (-32)}{1.8} = \frac{-72}{1.8} = -40°C$$

18. Work the exercises below.

How many of the indicated units are contained in the given unit?

a.   1 m = _____ cm   b.   1 m = _____ mm

c.   1 km = _____ m   d.   1 L = _____ mL

e.   1 kg = _____ g   f.   1 cm = _____ mm

Solve the following problems.

g.   12 m = _____ cm   h.   876 mL = _____ L

i.   6.6 kg = _____ g   j.   28 cc = _____ mL

k.   14.7 cm = _____ mm   l.   489 mg = _____ g

m.   429 mm = _____ m   n.   3.8 L = _____ mL

o.   28 g of water have a volume of _____ mL

p.   9 oz (liquid) = _____ mL   q.   84 g = _____ oz (weight)

r.   254 cm = _____ in   s.   10 mi = _____ km

t.   8.8 lb = _____ kg

u.   The formula for converting Fahrenheit to Celsius is _____ .

v.   The formula for converting Celsius to Fahrenheit is _____ .

w.   −4°F on the Celsius scale is _____ .

x.   20°C on the Fahrenheit scale is _____ .

17. a. 86°F
    b. 5°F
    c. 0°C
    d. −5°C

Part 9. Density, Specific Gravity, and Specific Heat

1. There are physical characteristics of a substance that help identify the substance. The first characteristic to be examined is called density.

Density is defined as mass per unit volume. Density is calculated by dividing the mass of an object by its volume. This is shown with the following equation.

$$\text{density } (d) = \frac{\text{mass}}{\text{volume}}$$

We can calculate the density of a solid, liquid, or gas. The density of a gas will be dealt with in a later unit. Note carefully the difference in units in the formulas of the density of a solid and liquid.

$$d \text{ (solid)} = \frac{g}{cm^3}$$

$$d \text{ (liquid)} = \frac{g}{mL}$$

Density problems can be solved using a four-step procedure.

Example: 33.9 grams of a solid substance has a volume of 3.00 cm³. What is the density of this substance?

Step 1.   Write the formula for density.

$$d = \frac{\text{mass}}{\text{volume}}$$

18. a. 100
    b. 1000
    c. 1000
    d. 1000
    e. 1000
    f. 10
    g. 1200
    h. 0.876
    i. 6600
    j. 28
    k. 147
    l. 0.489
    m. 0.429
    n. 3800
    o. 28
    p. 270
    q. 3
    r. 100
    s. 16
    t. 4
    u. $\dfrac{°F - 32}{1.8}$
    v. (1.8 X °C) + 32
    w. −20°C
    x. 68°F

Step 2.   List the data given in the problem.

mass = 33.9 g

volume = 3.00 cm^3

Step 3.   Substitute the data into the formula.

$$d = \frac{33.9 \text{ g}}{3.00 \text{ cm}^3}$$

Step 4.   Do the calculation using the concepts of significant digits and dimensional analysis. Use your calculator to divide the numbers.

$$d = 11.3 \text{ g/cm}^3$$

No units are canceled in this problem, and therefore all units must show up in the answer.

Here is another example: What is the density of a liquid if 5.0 mL has a mass of 2.8g?

Step 1. $d = \dfrac{\text{mass}}{\text{volume}}$

Step 2.   mass = 2.8 g

volume = 5.0 mL

Step 3. $d = \dfrac{2.8 \text{ g}}{5.0 \text{ mL}}$

Step 4. $d = 0.56$ g/mL

Try this problem using all four steps.

What is the density of a solid if 12.96 g has a volume of 6.00 cm^3?

Step 1.

Step 2.

Step 3.

Step 4.

---

2. Solve these density problems.

   a.   What is the density of copper if 40.19 g has a volume of 4.50 cm^3 ?

   b.   What is the density of mercury if 5.00 mL has a mass of 67.76 g?

   c.   What is the density of alcohol if 20 mL has a mass of 15.8 g?

   d.   What is the density of a 20 g piece of wood that has a volume of 50 cm^3 ?

1. Step 1.

$$d = \frac{\text{mass}}{\text{volume}}$$

Step 2.
mass = 12.96 g

volume = 6.00 cm^3

Step 3.

$$d = \frac{12.96 \text{ g}}{6.00 \text{ cm}^3}$$

Step 4.
$d = 2.16$ g/cm^3

3. Often there is a need to find the mass of an object given the density and the volume.

> *Example:* What is the mass of 15.00 cm^3 of gold? The density of gold is 19.30 g/cm^3 at 20°C. The density of a substance varies with temperature, so the temperature is always stated. The temperature will not enter into the calculations.

Step 1. Write the formula.

$$d = \frac{mass}{volume}$$

We are asked to solve for the mass so the formula must be rearranged.

$$volume \times d = \frac{mass}{\cancel{volume}} \times \cancel{volume}$$

or mass = volume $\times$ d

Step 2. volume = 15.00 cm^3

d = 19.30 g/cm^3

Step 3. mass = 15.00 cm$^3 \times \dfrac{19.30 \text{ g}}{cm^3}$

Step 4. mass = 15.00 $\cancel{cm^3} \times \dfrac{19.30 \text{ g}}{\cancel{cm^3}}$

mass = 15.00 $\times$ 19.30 g

mass = 289.5 g

Try this problem doing all steps.

What is the mass of 22.0 mL of oil that has a density of 0.920 g/mL at 25°C?

Step 1.

Step 2.

Step 3.

Step 4.

4. Solve these problems.

a. What is the mass of 20.0 mL of sulfuric acid if the density of the acid is 1.84 g/mL at 20° C?

b. What is the mass of 40.0 cm^3 of silver if the density of silver is 10.5 g/cm^3 at 20°C?

c. What is the mass of 100 mL of alcohol with a density of 0.79 g/mL at 20° C?

d. What is the volume of 30.0 g of oil that has a density of 0.850 g/mL at 20° C? (Rearrange the density equation to solve for volume.)

---

2. a. 8.93 g/cm^3
   b. 13.6 g/mL
   c. 0.79 g/mL
   d. 0.40 g/cm^3

---

3. Step 1.
   mass = volume $\times$ d

   Step 2.
   volume = 22.0 mL
   d = 0.920 g/mL

   Step 3.
   mass = 22.0 mL
   $\times$ 0.920 g/mL

   Step 4.
   mass = 20.2 g

5. Another characteristic used to identify substances is called specific gravity. The specific gravity of a solid or liquid is the density of the solid or liquid compared to the density of water at 4°C.

$$\text{specific gravity (sp gr)} = \frac{\text{density of solid or liquid}}{\text{density of water at } 4°C}$$

Example: What is the specific gravity of mercury if the density of mercury is 13.55 g/mL? The specific gravity of water at 4°C is 1.000 g/mL.

Step 1. $\text{sp gr} = \dfrac{d \text{ of mercury}}{d \text{ of water at } 4°C}$

Step 2. d of mercury = 13.55 g/mL
         d of water (4°C) = 1.000 g/mL

Step 3. $\text{sp gr} = \dfrac{13.55 \text{ g/mL}}{1.000 \text{ g/mL}}$

Step 4.   sp gr = 13.55

      All units cancel: specific gravity is a ratio.
      The number 13.55 means that mercury is 13.55 times heavier than water.

Calculate the specific gravity of the following. The density of water is always 1.000 g/mL at 4° C.

a.   The density of gasoline is 0.56 g/mL. What is its specific gravity?

b.   The density of a sugar solution is 1.15 g/mL. What is its specific gravity?

6. Solve the following problems using the concepts of significant digits and dimensional analysis. Use your calculator to perform the computations.

a.   13.6 g of a substance has a volume of 15.0 mL. What is the density of this substance?

b.   What is the density of a substance if 200 cm³ weighs 250 g?

c.   What is the mass of 18.0 cm³ of brass if the density of brass is 8.00 g/cm³?

d.   What is the mass of 25.5 mL of an acid solution that has a density of 1.19 g/mL?

e.   First calculate the density and then find the specific gravity of a solid that weighs 180 g and has a volume of 40.0 cm³.

f.   What is the mass of 30.0 mL of a liquid that has a specific gravity of 0.850?

7. Answers to frame 6.

    a.   0.907 g/mL   b.   1.25 g/cm³   c.   144 g
    d.   30.3 g   e.   d = 4.50 g/cm³ and sp gr = 4.50   f.   25.5 g

---

4. a. 36.8 g
   b. 420 g
   c. 79 g
   d. 35.3 mL

5. a. 0.56
   b. 1.15

NAME _____

THE METRIC SYSTEM, DENSITY
AND SPECIFIC GRAVITY
EVALUATION TEST 1
PART 1-PART 3

How many of these subdivisions are in a standard unit?

1.   milli   2.   centi   3.   deci

How many of the standard units are in each of these multiples?

4.   deka   5.   kilo   6.   hecto

How many of the indicated units are contained in the given unit?

7.   In one kilometer there are _____ m.

8.   In one meter there are _____ cm.

9.   In one meter there are _____ mm.

10. In one centimeter there are _____ mm.

Set up, but do not solve, the following conversion equations.

11.   Change 10 m to centimeters.

12.   Change 89 m to kilometers.

13.   Change 19 cm to millimeters.

Solve these problems.

14.   Change 1.86 m to millimeters.

15.   Change 192 cm to meters.

16.   Change 150 mm to meters.

17.   Change 0.51 m to centimeters.

18.   Change 13 mm to centimeters.

1. _____

2. _____

3. _____

4. _____

5. _____

6. _____

7. _____

8. _____

9. _____

10. _____

11. _____

12. _____

13. _____

14. _____

15. _____

16. _____

17. _____

18. _____

NAME _____

THE METRIC SYSTEM, DENSITY
AND SPECIFIC GRAVITY
EVALUATION TEST 2
PART 4-PART 7

How many of the indicated units are contained in the given unit?

1. In 1 liter there are _____ mL.

2. In 10 mL there are _____ cm³.

3. In 100 cc there are _____ mL.

4. In 1 g there are _____ mg.

5. In 1 g there are _____ cg.

6. In 1 kg there are _____ g.

Solve the following, using conversion equations.

7. Change 798 mL to liters

8. Change 3.6 liters to milliliters.

9. Change 92 mL to cubic centimeters.

10. Change 4.4 kg to grams.

11. Change 429 g to kilograms.

12. Change 5.6 g to milligrams.

13. Mass is the _____ an object possesses.

14. Weight is caused by _____ .

15. Mass is measured on a _____ .

16. 10 g of water have a volume of _____ mL.

17. 28 cc of water have a mass of _____ g.

18. Change 908 g to pounds.

19. Change 10.0 m to inches.

20. Change 6.6 lb to kilograms.

21. Change 5.00 qt to liters.

1. _____

2. _____

3. _____

4. _____

5. _____

6. _____

7. _____

8. _____

9. _____

10. _____

11. _____

12. _____

13. _____

_____

14. _____

15. _____

16. _____

17. _____

18. _____

19. _____

20. _____

21. _____

NAME _____

THE METRIC SYSTEM, DENSITY
AND SPECIFIC GRAVITY
EVALUATION TEST 3
PART 8

1. The boiling point of water is _____ °F.

2. The freezing point of water is _____ °C.

3. 212°F is equal to _____ °C.

4. 0°C is equal to _____ °F.

5. Which is larger—a degree Celsius or a degree Fahrenheit?

6. Write the formula for converting Fahrenheit to Celsius.

7. Write the formula for converting Celsius to Fahrenheit.

8. What is 30°C on the Fahrenheit scale?

9. What is 41°F on the Celsius scale?

10. What is −10°C on the Fahrenheit scale?

11. What is 16°F on the Celsius scale?

1. _____

2. _____

3. _____

4. _____

5. _____

6. _____

7. _____

8. _____

9. _____

10. _____

11. _____

NAME _____

THE METRIC SYSTEM, DENSITY
AND SPECIFIC GRAVITY
EVALUATION TEST 4
PART 9

1. 420 g of a substance has a volume of 250 cm^3. What is the density of this substance?

2. What is the density of a substance if 30.5 mL weighs 28.6 g?

3. What is the mass of 11.6 cm^3 of gold if the density of gold is 19.3 g/cm^3 ?

4. What is the mass of 20.40 mL of mercury if the density of mercury is 13.55 g/mL?

5. Calculate the density and find the specific gravity of 150 mL of a liquid that weighs 115 g.

6. What is the mass of a solid that has a volume of 27.0 cm^3 and a specific gravity of 7.78?

1. _____

2. _____

3. _____

4. _____

5. _____

6. _____

erdict /PMpub` save put

UNIT 6
The Division and Properties of Matter

In this unit you will learn to classify matter. The terms matter, element, atom, compound, *and* molecule *will be defined. You will learn to identify the various states of matter and to differentiate between chemical and physical changes in matter.*

Part 1. Matter

1. *Matter* is anything that has mass and occupies space. Examples of matter are water, wood, and iron. Not all forms of matter are visible, however. Air is an example of invisible matter. If a glass is submerged mouth downward in a container of water, very little water will rise into the glass. Air, which is matter, occupies the space within the glass and keeps the water from getting in.

   Indicate which of the following are matter. _____

   a. milk   b. wood   c. heat   d. oxygen gas

2. Food is matter because it (a) _____ space and has (b) _____ .

3. a. Is electricity matter? _____

   b. Why? _____

1. a, b, and d

2. a. occupies
   b. mass

3. a. no
   b. It does not have mass or occupy space.

Part 2. Elements

1. An object may consist of many types of matter or only one type of matter. When an object consists of only one type of matter, it is called a *pure substance*. One form of a pure substance is the element.

2. An element consists of only _____ type of matter.

3. Because an element consists of only one type of matter, it is a _____

   _____ .

2. one

3. pure substance

4. The following table contains a list of the most common elements. Most of the symbols for the elements are derived from the modern names. Where the symbol is taken from the ancient name, the ancient name is given to help you learn the symbol. You need not learn the ancient name of the element. You must know the modern names, with correct spelling, and the symbols for these elements.

Names and Symbols of the Most Common Elements

The first letter of the symbol is always capitalized. The second letter is never capitalized.

Element	Symbol	Element	Symbol
aluminum	Al	magnesium	Mg
barium	Ba	mercury (*hydrargyrum*)	Hg
boron	B	neon	Ne
bromine	Br	nitrogen	N
calcium	Ca	oxygen	O
carbon	C	phosphorus	P
chlorine	Cl	potassium (*kalium*)	K
chromium	Cr	radium	Ra
copper (*cuprum*)	Cu	silicon	Si
fluorine	F	silver (*argentum*)	Ag
gold (*aurum*)	Au	sodium (*natrium*)	Na
helium	He	sulfur	S
hydrogen	H	tin (*stannum*)	Sn
iodine	I	titanium	Ti
iron (*ferrum*)	Fe	uranium	U
lead (*plumbum*)	Pb	zinc	Zn

Part 3. Atoms

1. The smallest particle of an element that can exist and still retain the properties of the element is the atom. All matter is made up of these tiny particles. At this stage we will assume that all atoms of the same element are alike. Later we will learn that there are some differences.

Example: A bar of gold is made up of a vast number of gold atoms.

When you hold a bar of iron in your hand, you are holding a large number of _____

_____ .

2. If you have an object composed only of silver atoms, you have the element _____

_____ .

1. iron atoms

3. A gold atom in one bar of gold is exactly like a gold atom in any other bar of gold. A gold atom, however, is different from an atom of any other element.

2. silver

Which two objects below have the same kind of atoms? _____

a. iron fork

b. silver fork

c. aluminum spoon

d. silver spoon

4. Why is it that a silver fork and a silver spoon have the same kind of atoms but a silver fork and an iron fork have different kinds of atoms? _____

_____

5. Which of the objects below would have atoms different from those of the other three? _____

a. silver quarter   b. silver dime

c. lead nickel   d. silver dollar

Part 4. Compounds

1. Compounds are formed when the atoms of two or more elements combine chemically. Water, for example, is a compound made up of hydrogen and oxygen atoms.

When atoms of two or more elements combine chemically, they form _____

_____ .

2. Compounds are also pure substances, because all of the particles of a compound are alike. These particles of compounds are called *molecules*.

Because all of the particles of a compound are alike, a compound is a _____

_____ .

Part 5. Molecules

1. We have seen that the smallest particle of an element that can exist is the atom. One of the smallest particles of a compound that can exist is the molecule. (Compounds can also be composed of other small particles, such as ions, but for now we will discuss only the molecule.) A molecule is to a compound as an atom is to an element. A *molecule* is defined as two or more atoms that are chemically united.

Elements are composed of atoms and compounds are composed of _____ .

2. When the atoms of two or more elements are chemically united, we have a _____

_____ of a compound.

3. If you have a vast number of water molecules in a glass, you have the compound called _____ .

4. Molecules of the same compound are all alike. Molecules of different compounds are different.

In which of the examples below are the molecules all alike? _____

a. a glass of water   b. a glass of alcohol

c. a cup of water   d. a cup of gasoline

Answer column (right margin):

3. b and d

4. Because elements of the same atom are alike.

5. c

1. compounds

2. pure substance

1. molecules

2. molecule

3. water

5. Indicate which of the following items would have molecules different from the other three. _____

   a. cube of sugar

   b. spoonful of sugar

   c. stick of sugar

   d. spoonful of sand

4. a and c

Part 6. The States of Matter

1. There are three states of matter: solid, liquid, and gas. In the solid state, matter has a definite shape and a definite volume. The desk you sit at and the car you drive have the same shape and occupy the same volume today as they did yesterday because they are solids.

    When you use a fork at your next meal, you will have no trouble recognizing it as a solid because it will have (a) _____ shape and (b) _____ volume.

5. d

2. In the liquid state, matter has an indefinite shape but a definite volume. A liquid takes the shape of the container it is in but the volume remains the same. If we poured a quart of milk, for example, into each of the containers shown below, the shape of the quart of milk would be different in each case but the volume would be the same.

1. a. a definite
   b. a definite

   a. If you pour a cup of water into a glass bottle, does the volume of water change? _____

   b. Does the shape of the water change? _____

3. Liquids and solids have in common the characteristic that they both have a definite _____ .

2. a. no
   b. yes

4. A gas has an indefinite shape and an indefinite volume. A gas will assume the shape and volume of the container it is in. If you have a balloon filled with helium gas, the gas has the shape and volume of the balloon. If you let the gas out of the balloon, the gas will take the shape and volume of the room you are in.

    A gas is most easily identified by noting that it has an (a) _____ shape and an (b) _____ volume.

3. volume

5. A liquid and a gas have a characteristic in common. They both have an indefinite _____ .

4. a. indefinite
   b. indefinite

6. Indicate which of the following choices states the characteristics of a solid. _____

    a.   indefinite shape, definite volume
    b.   definite shape, definite volume
    c.   indefinite shape, indefinite volume

7. Indicate which of the following substances would have an indefinite shape and an indefinite volume. _____

    a.   liquid   b. solid   c.   gas

Part 7. Chemical and Physical Changes in Matter

1. Matter can undergo two different kinds of changes: physical and chemical. A change in the size, shape, or state of matter is called a *physical change*. During a physical change no new substances are formed. Water, for example, can exist as a solid (ice), liquid, or gas (steam), but in all of these states it is still water. The size of a block of ice can vary, but the substance of which the ice is formed remains the same. The molecules making up the block of ice would still be water molecules.

   There are three types of physical changes that can occur. Change in size, change in

   (a) _____ , and change in (b) _____ .

2. In a chemical change a *new substance* is formed that is entirely different from the starting material. When a piece of iron rusts, the rust that forms is a substance entirely different from the original iron.

   When a chemical change occurs, an entirely _____ is formed.

3. Which of the following is a physical change? _____

    a.   melting of lead   b.   digesting of food   c.   burning of wood

4. Which of the following is a chemical change? _____

    a.   sawing of wood   b.   breaking of glass   c.   burning of gasoline

5. Here is a summary of the divisions of matter.

Matter
(Has mass and occupies space.)

Pure substance
(Composed of only
one kind of matter.)

Mixture
(Composed of several
kinds of matter.)

Elements
(Composed of only
one kind of atom.)

Compound
(Composed of only one
kind of molecule.)

Atoms of the elements
combine to form   ————▶ *Molecules* of compounds

5. shape

6. b

7. c

1. a. shape
   b. state

2. new substance

3. a

4. c

6. Give the names of the following elements, spelled correctly.

a. Zn _____   b. S _____

c. Si _____   d. O _____

e. Mg _____   f. H _____

g. Cr _____   h. Ca _____

i. Al _____   j. Pb _____

k. Ag _____   l. Hg _____

m. F _____   n. Sn _____

o. Cl _____   p. P _____

q. C _____   r. Au _____

s. He _____   t. U _____

u. K _____   v. Ba _____

w. I _____   x. N _____

y. Na _____   z. Br _____

aa. Cu _____   bb. B _____

cc. Fe _____   dd. Ti _____

7. Give the symbols of the following elements.

a. hydrogen _____   b. tin _____   c. zinc _____

d. oxygen _____   e. calcium _____   f. fluorine _____

g. titanium _____   h. potassium _____   i. magnesium _____

j. chlorine _____   k. chromium _____   l. neon _____

m. silicon _____   n. barium _____   o. helium _____

p. lead _____   q. iron _____   r. sodium _____

s. nitrogen _____   t. boron _____   u. sulfur _____

v. radium _____   w. carbon _____   x. aluminum _____

y. copper _____   z. iodine _____   aa. bromine _____

bb. phosphorus _____   cc. mercury _____   dd. silver _____

6. a. zinc
   b. sulfur
   c. silicon
   d. oxygen
   e. magnesium
   f. hydrogen
   g. chromium
   h. calcium
   i. aluminum
   j. lead
   k. silver
   l. mercury
   m. fluorine
   n. tin
   o. chlorine
   p. phosphorus
   q. carbon
   r. gold
   s. helium
   t. uranium
   u. potassium
   v. barium
   w. iodine
   x. nitrogen
   y. sodium
   z. bromine
   aa. copper
   bb. boron
   cc. iron
   dd. titanium

8. Fill in the word or words that best fit these definitions.

a.   A change in the size, shape, or state of matter. _____

b.   Matter formed by atoms that chemically unite. _____

c.   A state of matter characterized by a definite shape and volume. _____

d.   Has mass and occupies space. _____

e.   Only one kind of these is contained in any element. _____

f.   The smallest particle of a compound. _____

g.   A change in matter that results in the formation of a new substance.

_____

h.   Only one kind of this form of matter is contained in a compound.

_____

i.   The melting of ice is an example of this. _____

j.   A state of matter characterized by a definite volume and an indefinite shape.

_____

7.	a. H	b. Sn
	c. Zn	d. O
	e. Ca	f. F
	g. Ti	h. K
	i. Mg	j. Cl
	k. Cr	l. Ne
	m. Si	n. Ba
	o. He	p. Pb
	q. Fe	r. Na
	s. N	t. B
	u. S	v. Ra
	w. C	x. Al
	y. Cu	z. I
	aa. Br	bb. P
	cc. Hg	dd. Ag

9. Answers to frame 8.

a. physical change	b. molecules	c. solid	d. matter
e. atom	f. molecule	g. chemical change	
h. molecule	i. physical change	j. liquid	

THE DIVISION AND PROPERTIES
OF MATTER
EVALUATION TEST
PART 1–PART 7

Supply the terms that best answer the following questions.

1. What has mass and occupies space?

2. What form of matter contains only one kind of atom?

3. What is the smallest particle of an element?

4. What is the smallest particle of a compound?

5. What form of matter results when two or more elements chemically combine?

6. What state of matter has a definite shape and volume?

7. What state of matter has no definite shape or volume?

8. What kind of change takes place when the size, shape, or state of matter is altered?

9. What kind of change takes place when a new substance is formed from an original one?

10. What form of matter results when two or more atoms chemically combine?

State whether the following changes in matter are physical or chemical changes.

11.   melting of ice

12.   rusting of iron

Give the symbols for the following elements.

13. boron	14. chlorine
15. gold	16. mercury
17. zinc	18. calcium
19. sulfur	20. chromium
21. iodine	

Give the name for the following symbols.

22. Mg	23. Cu
24. K	25. Ag
26. Sn	27. P
28. He	29. Pb
30. C	

1. _____

2. _____

3. _____

4. _____

5. _____

6. _____

7. _____

8. _____

9. _____

10. _____

11. _____

12. _____

13. _____

14. _____

15. _____

16. _____

17. _____

18. _____

19. _____

20. _____

21. _____

22. _____

23. _____

24. _____

25. _____

26. _____

27. _____

28. _____

29. _____

30. _____

UNIT 7
Atomic Structure

In this unit you will learn the basic structure of the atom. The major subatomic particles will be identified and located within the atom. You will learn the arrangement of the electron-dot symbols, showing the number of electrons in the outer level of an atom. You will also learn what an ion is and how it is formed and how to relate atomic structure to the periodic table of the elements. You will learn to write the arrangement of electrons in the energy sublevels of the atom.

Part 1. Subatomic Particles

1. The major subatomic particles of an atom are the proton, neutron, and electron. It is convenient to give these particles the symbols shown in the table below.

Particle	Symbol
electron	e^-
proton	p
neutron	n

Which is the correct symbol for the proton? _____

a. n b. p c. e^-

2. The symbol e^- stands for the _____ .

1. b

3. Two of these subatomic particles contain an electrical force. This force is most often referred to as an *electrical charge*. There are two types of electrical forces, or charges: positive charges and negative charges.

Another name for a positive electrical force is a positive _____ .

2. electron

4. Charges can be positive or _____ .

3. charge

5. These electrical charges act as attracting or repelling forces. They attract or repel in the following manner:

$$\text{positive} \longrightarrow \text{(attract)} \longleftarrow \text{negative}$$
$$\text{positive} \longleftarrow \text{(repel)} \longrightarrow \text{positive}$$
$$\text{negative} \longleftarrow \text{(repel)} \longrightarrow \text{negative}$$

Thus unlike charges attract and like charges repel.

In which of these situations would there be an attraction? _____

a.   positive–positive   b.   positive–negative   c.   negative–negative

6. Two negative charges would _____ each other.

5. b

7. In addition to the sign of the charge (positive or negative), we can also talk about the quantity of charge a subatomic particle has. For example, the proton has one unit of positive charge and the electron has one unit of negative charge. Because protons and electrons are oppositely charged, they will attract each other. Protons will repel other protons because they are charged alike, and electrons will repel other electrons because they are charged alike. Neutrons have no charge and neither repel nor attract. The table below shows the charges of these subatomic particles.

6. repel

Particle	Symbol	Charge
proton	p	+1
electron	e^-	−1
neutron	n	0

In which of these situations would there be an attraction? _____

a.   proton–proton   b.   electron–electron   c.   electron–proton

8. In which of these situations would the particles repel each other? _____

a.   two electrons   b.   electron and proton

7. c

9. The proton has the symbol (a) _____ and a charge of (b) _____ .

8. a

10. The particle that has a charge of −1 is called the (a) _____ and has the symbol (b) _____ .

9. a. p
   b. +1

11. The particle that has the symbol n is called the (a) _____ and has a charge of (b) _____ .

10. a. electron
    b. e^-

Part 2. Location of Subatomic Particles in the Atom

1. The two major areas of the atom are the nucleus and the energy levels that surround it. The nucleus is the extremely dense and tiny center of the atom that contains the protons and neutrons. The energy levels contain all of the electrons. Electrons arrange themselves in levels according to the amount of energy associated with that level. The diagram below is a representation of the arrangement of the atom.

11. a. neutron
    b. 0

Nucleus (containing protons and neutrons)

Energy levels (containing electrons)

The two major areas of the atom are the (a) _____ and the

(b) _____ .

_____

2. The nucleus contains neutrons and _____ .

1. a. nucleus
   b. energy levels

_____

3. The energy levels contain the _____ .

2. protons

_____

4. Because the nucleus contains the (a) _____ , it has a

(b) _____ charge.

3. electrons

_____

5. The energy levels have a negative charge, because they contain the

_____ .

4. a. protons
   b. positive

_____

6. The neutrons are in the (a) _____ and have a charge of

(b) _____ .

5. electrons

_____

Periodic Table of the Elements

Atomic masses are rounded off to the nearest 0.1.
[a] Mass number of most stable or best-known isotope.
[b] Mass of most commonly available long-lived isotope.

Noble Gases

Group IA																	Noble Gases
1 Hydrogen **H** 1.0	IIA											IIIA	IVA	VA	VIA	VIIA	2 Helium **He** 4.0
3 Lithium **Li** 6.9	4 Beryllium **Be** 9.0											5 Boron **B** 10.8	6 Carbon **C** 12.0	7 Nitrogen **N** 14.0	8 Oxygen **O** 16.0	9 Fluorine **F** 19.0	10 Neon **Ne** 20.2
11 Sodium **Na** 23.0	12 Magnesium **Mg** 24.3	IIIB	IVB	VB	VIB	VIIB	VIII			IB	IIB	13 Aluminum **Al** 27.0	14 Silicon **Si** 28.1	15 Phosphorus **P** 31.0	16 Sulfur **S** 32.1	17 Chlorine **Cl** 35.5	18 Argon **Ar** 39.9
19 Potassium **K** 39.1	20 Calcium **Ca** 40.1	21 Scandium **Sc** 45.0	22 Titanium **Ti** 47.9	23 Vanadium **V** 50.9	24 Chromium **Cr** 52.0	25 Manganese **Mn** 54.9	26 Iron **Fe** 55.8	27 Cobalt **Co** 58.9	28 Nickel **Ni** 58.7	29 Copper **Cu** 63.5	30 Zinc **Zn** 65.4	31 Gallium **Ga** 69.7	32 Germanium **Ge** 72.6	33 Arsenic **As** 74.9	34 Selenium **Se** 79.0	35 Bromine **Br** 79.9	36 Krypton **Kr** 83.8
37 Rubidium **Rb** 85.5	38 Strontium **Sr** 87.6	39 Yttrium **Y** 88.9	40 Zirconium **Zr** 91.2	41 Niobium **Nb** 92.9	42 Molybdenum **Mo** 95.9	43 Technetium **Tc** 98.9[b]	44 Ruthenium **Ru** 101.0	45 Rhodium **Rh** 102.9	46 Palladium **Pd** 106.4	47 Silver **Ag** 107.9	48 Cadmium **Cd** 112.4	49 Indium **In** 114.8	50 Tin **Sn** 118.7	51 Antimony **Sb** 121.8	52 Tellurium **Te** 127.6	53 Iodine **I** 126.9	54 Xenon **Xe** 131.3
55 Cesium **Cs** 132.9	56 Barium **Ba** 137.3	57 Lanthanum **La** 138.9	72 Hafnium **Hf** 178.5	73 Tantalum **Ta** 180.9	74 Tungsten **W** 183.8	75 Rhenium **Re** 186.2	76 Osmium **Os** 190.2	77 Iridium **Ir** 192.2	78 Platinum **Pt** 195.1	79 Gold **Au** 197.0	80 Mercury **Hg** 200.6	81 Thallium **Tl** 204.4	82 Lead **Pb** 207.2	83 Bismuth **Bi** 208.9	84 Polonium **Po** (210)[a]	85 Astatine **At** (210)[a]	86 Radon **Rn** (222)[a]
87 Francium **Fr** (223)[a]	88 Radium **Ra** 226.0[b]	89 Actinium** **Ac** (227)[a]	104 Unnilquadium **Unq** (261)[a]	105 Unnilpentium **Unp** (262)[a]	106 Unnilhexium **Unh** (263)[a]	107 Unnilseptium **Uns** (262)[a]	108 Unniloctium **Uno** (265)[a]	109 Unnilennium **Une** (266)[a]									

*Lanthanide Series 6

58 Cerium **Ce** 140.1	59 Praseodymium **Pr** 140.9	60 Neodymium **Nd** 144.2	61 Promethium **Pm** (145)[a]	62 Samarium **Sm** 150.4	63 Europium **Eu** 152.0	64 Gadolinium **Gd** 157.2	65 Terbium **Tb** 158.9	66 Dysprosium **Dy** 162.5	67 Holmium **Ho** 164.9	68 Erbium **Er** 167.3	69 Thulium **Tm** 168.9	70 Ytterbium **Yb** 173.0	71 Lutetium **Lu** 175.0

** Actinide Series 7

90 Thorium **Th** 232.0[b]	91 Protactinium **Pa** 231.0[b]	92 Uranium **U** 238.0	93 Neptunium **Np** 237.0[b]	94 Plutonium **Pu** (242)[a]	95 Americium **Am** (243)[a]	96 Curium **Cm** (247)[a]	97 Berkelium **Bk** (249)[a]	98 Californium **Cf** (251)[a]	99 Einsteinium **Es** (254)[a]	100 Fermium **Fm** (253)[a]	101 Mendelevium **Md** (256)[a]	102 Nobelium **No** (254)[a]	103 Lawrencium **Lr** (257)[a]

Key

Atomic Number → 11
Name → Sodium
Symbol → **Na**
Atomic Mass → 23.0

Part 3. Electron Structure of the Atom

1. The number and arrangement of electrons in the energy levels of the atom determine what type of compounds the element will form. To determine the number of electrons in an atom, we consult the periodic chart of the elements on page 114. (The periodic chart is also reproduced on page 338.) The chart lists the atomic number, symbol, and atomic weight of each element. The element potassium, for example, is listed as follows.

<table>
<tr><td rowspan="4">19
Potas-
sium
K
39.1</td><td>← atomic number</td></tr>
<tr><td>← name</td></tr>
<tr><td>← symbol</td></tr>
<tr><td>← atomic weight</td></tr>
</table>

The atomic number represents the number of protons in the atom. (It is not necessary to discuss the atomic weight at this time.)

Use the periodic chart to find the name and symbol for the following elements.

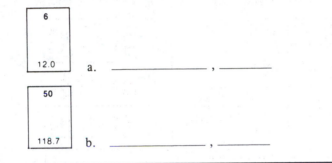

a. _____ , _____

b. _____ , _____

6. a. nucleus
   b. zero

2. Use the periodic chart to find the atomic numbers of the following elements.

Iron **Fe** 55.8	Gold **Au** 197.0

a. _____   b. _____

1. a. carbon, C
   b. tin, Sn

3. The atomic number tells how many _____ are present in the atom.

2. a. 26
   b. 79

4. If an element has an atomic number of 79, there are _____ protons in an atom of that element.

3. protons

5. All atoms are neutral. This means that there is always an equal number of positive and negative charges in an atom. Because protons are positively charged and electrons are negatively charged, there must therefore be an equal number of electrons and protons in an atom.

An element with 29 electrons per atom must also have _____ protons per atom.

4. 79

6. The atomic number represents the number of protons in an atom, which is also the number of electrons since there are an equal number of protons and electrons in every atom.

Example:

Element	Atomic Number	Number of Protons	Number of Electrons
Ne	10	10	10

Fill in the needed information below.

Element	Atomic Number	Number of Protons	Number of Electrons
Cl	a. _____	b. _____	c. _____

7. Because the number of protons and electrons are the same in an atom, we can deduce the number of electrons directly from the atomic number.

Example:

Element	Atomic Number	Number of Electrons
Fe	26	26

Fill in the atomic number and the number of electrons.

Element	Atomic Number	Number of Electrons
Sc	a. _____	b. _____

8. The next task is to determine how many electrons go into each of the seven energy levels. We can do this by using the mathematical formula $2n^2$, where n is the number of the energy level.

Example:

Energy Level = n	Substitute for n in Formula	Maximum Number of Electrons in Level
1	$2n^2 = 2(1)^2$	$2(1 \times 1) = 2$
2	$2n^2 = 2(2)^2$	$2(2 \times 2) = 8$

Below, determine the maximum number of electrons in the third and fourth energy levels.

Energy Level = n	Substitute for n in Formula	Maximum Number of Electrons in Level
3	a. _____	b. _____
4	c. _____	d. _____

5. 29

6. a. 17
    b. 17
    c. 17

7. a. 21
    b. 21

9. To arrange the electrons in the energy levels, we use our ability to find the total number of electrons an atom has and our ability to find the maximum number of electrons in each energy level. Except for a limitation that will be explained later, the energy levels fill from the first one on out. The first level of any atom can hold a maximum of two electrons and the second level can hold a maximum of eight.

Example:

Element	Number of Electrons	Arrangement of Electrons in Energy Levels 1 2 3 4
F	9	2 7

The fluorine atom has a total of nine electrons to distribute in the seven levels. The first level took two of the nine electrons, leaving only seven for the second level even though it could hold eight.

Complete the following.

Element	Number of Electrons	Arrangement of Electrons in Energy Levels 1 2 3 4
S	a. _____	b. _____

Note that although the third level of an atom can hold 18 electrons, we have only 6 left to go there in the sulfur atom.

10. Determine the electron arrangement of these elements.

Element	Number of Electrons	Arrangement of Electrons in Energy Levels 1 2 3 4
P	a. _____	b. _____
Mg	c. _____	d. _____

11. There is a further limitation on the number of electrons in an energy level known as *the rule of eight*. This rule states that no *outer* level (the last level containing electrons) may contain more than eight electrons. The rule of eight does not apply to levels one and two because level one can hold a maximum of only two electrons and level two can hold no more than eight electrons.

Example:

Element	Number of Electrons	Arrangement of Electrons in Energy Levels 1 2 3 4
Ca	20	2 8 8 2

We could not put 10 electrons in the third level, even though it can hold a maximum of 18, because then it would be the outermost level containing electrons and the rule of eight would apply.

8. a. $2n^2 = 2(3)^2$
   b. $2(3 \times 3) = 18$
   c. $2n^2 = 2(4)^2$
   d. $2(4 \times 4) = 32$

9. a. 16
   b. 2, 8, 6, 0

10. a. 15
    b. 2, 8, 5, 0
    c. 12
    d. 2, 8, 2, 0

Find the electron arrangement of this element.

Element	Number of Electrons	Arrangement of Electrons in Energy Levels 1 2 3 4
K	a. _____	b. _____

---

12. Levels three through six can hold more than eight electrons if they are not the outermost level containing electrons. Electrons can be added to an inner level if two electrons are first added to the next outermost level.

11. a. 19
    b. 2, 8, 8, 1

Example:

Element	Number of Electrons	Arrangement of Electrons in Energy Levels 1 2 3 4
Sc	21	2 8 9 2

Because there are two electrons in the fourth level, the third level can hold more than 8 electrons, up to its maximum of 18 electrons.

Example:

Element	Number of Electrons	Arrangement of Electrons in Energy Levels 1 2 3 4
Zn	30	2 8 18 2

Because there are two electrons in the outermost level, a maximum of 18 electrons can now be put in level three.

Find the electron arrangement of this element.

Element	Number of Electrons	Arrangement of Electrons in Energy Levels 1 2 3 4
Mn	a. _____	b. _____

---

13. Find the electron arrangements of these elements.

12. a. 25
    b. 2, 8, 13, 2

Element	Number of Electrons	Arrangement of Electrons in Energy Levels 1 2 3 4
Ni	a. _____	b. _____
Ti	c. _____	d. _____

Part 4. Electron-Dot Symbols

1. The symbol for an element together with dots representing the number of electrons in the outermost level is called an *electron-dot symbol*.

   Example:

   Na• = sodium, which has one electron in its outermost level

   :C̈l: = chlorine, which has seven electrons in its outermost energy level

   How many electrons are in the outer energy levels of these elements?

   a. Ca: _____   b. :N̈• _____

2. To determine the electron-dot symbol, the arrangement of electrons in the energy levels must be known.

   Example:

Element	Number of Electrons	Arrangement of Electrons in Energy Levels 1 2 3 4
Mg	12	2 8 2

   In the element magnesium the number of electrons in the last level is two. The electron-dot symbol is (Mg:).

   The electron-dot symbols for the most common elements are listed below.

   H•   He:

   Li•   Be: :B̈   :C̈•   :N̈•   •Ö: :F̈:   :N̈e:

   Na•   Mg: :Äl   :S̈i•   :P̈•   •S̈:   :C̈l: :Är:

   K•   Ca:

   Note that in most cases the electrons will form pairs.

   a. Write the electron-dot symbol for the following element. _____

Element	Number of Electrons	Arrangement of Electrons in Energy Levels 1 2 3 4
P	15	2 8 5

   b. Number of electrons in the last level? _____

3. Write the electron-dot symbols for the following elements.

   a. Na _____   b. K _____

   c. Mg _____   d. Ca _____

   e. N _____   f. P _____

   g. F _____   h. Cl _____

---

13. a. 28
    b. 2, 8, 16, 2
    c. 22
    d. 2, 8, 10, 2

1. a. 2
   b. 5

2. a. :P̈•
   b. 5

Part 5. Ion Formation

1. Ions are atoms or groups of atoms that have lost or gained electrons. When an atom has more electrons than protons, or vice versa, it becomes an ion. A sodium atom, for example, has 11 protons and 11 electrons. A sodium ion has 11 protons but only 10 electrons. This information is summarized in the following table.

	Protons		Electrons		Charge on Atom or Ion	Symbol
sodium atom	(+11)	+	(−11)	=	0	Na
sodium ion	(+11)	+	(−10)	=	+1	Na^{1+}

Note that the +1 charge shows up in the symbol for the ion.

Write the symbol for this ion, showing the correct charge.

	Protons		Electrons		Charge on Ion	Symbol and Charge
Mg ion	(+12)	+	(−10)		a. _____	b. _____

2. Use your periodic table to find the symbol for this ion and write the symbol with the correct charge. Keep in mind that the number of protons is the atomic number.

Protons		Electrons		Charge on Ion	Symbol and Charge
(+17)	+	(−18)		a. _____	b. _____

3. Use your periodic table to find the symbols for these ions and write the symbol, showing the correct charge.

Protons		Electrons		Symbol and Charge
(+11)	+	(−10)		a. _____
(+8)	+	(−10)		b. _____
(+16)	+	(−18)		c. _____
(+13)	+	(−10)		d. _____

4. To understand why atoms lose or gain electrons to form ions, we must look at the number of electrons in the outermost energy level. Most elements seek to have eight electrons in the outermost energy level. To accomplish this, they gain or lose electrons in the following manner.

Example:	Electron Arrangement 1 2 3	Symbol and Charge
sodium atom	2 8 1	Na (no charge)

If sodium could lose the one electron in its third energy level, the second level would be the outermost level and would contain eight electrons.

Answer column (right side):

3. a. Na·
   b. K·
   c. Mg:
   d. Ca:
   e. :N̈·
   f. :P̈·
   g. :F̈:
   h. :C̈l:

1. a. +2
   b. Mg^{2+}

2. a. −1
   b. Cl^{1-}

3. a. Na^{1+}
   b. O^{2-}
   c. S^{2-}
   d. Al^{3+}

	Electron Arrangement 1 2 3	Symbol and Charge
sodium ion	2 8	Na^{1+}

When the sodium atom loses one electron, it forms the sodium ion with a charge of +1. (Keep in mind that electrons have −1 charge and that when you subtract a −1 you get a +1.)

Here is a table showing how three atoms lose electrons to form ions. Note that after losing the electrons each element will have eight electrons in its outer energy level.

Element	Electron Arrangement		Ion Formed
Na	2 8 1	(lose one)	Na^{1+}
Mg	2 8 2	(lose two)	Mg^{2+}
Al	2 8 3	(lose three)	Al^{3+}

Write the electron arrangement, state the number of electrons lost, and give the symbol of the ion for the following element.

Element	Electron Arrangement	Number of Electrons Lost	Symbol for Ion
Ca	a. _____	b. _____	c. _____

---

5. Indicate the electron arrangement, state the number of electrons lost, and give the symbol of the ions for these elements.

Element	Electron Arrangement	Number of Electrons Lost	Symbol for Ion
B	a. _____	b. _____	c. _____
K	d. _____	e. _____	f. _____

---

6. When elements have one, two, or three electrons in the outer energy level, they will lose these electrons to form ions. When they have five, six, or seven electrons in the outer energy level, they will often gain electrons to form ions. Examples of this are shown below.

Element	Electron Arrangement		Ion Formed
N	2 5	(gain three)	N^{3-}
O	2 6	(gain two)	O^{2-}
F	2 7	(gain one)	F^{1-}

Note that, when these elements gain electrons, they have eight electrons in the outer energy level. When electrons are gained, the ion is negatively charged.

Write the electron arrangement, state how many electrons were gained, and give the symbol of the ion for the following element.

Element	Electron Arrangement	Number of Electrons Gained	Symbol for Ion
S	a. _____	b. _____	c. _____

4. a. 2 8 8 2
   b. 2
   c. Ca^{2+}

5. a. 2 3
   b. 3 lost
   c. B^{3+}
   d. 2 8 8 1
   e. 1 lost
   f. K^{1+}

7. Write the electron arrangement, state the number of electrons gained, and give the symbol of the ion formed for these elements.

Element	Electron Arrangement	Number of Electrons Gained	Symbol for Ion
P	a. _____	b. _____	c. _____
Cl	d. _____	e. _____	f. _____

Part 6. Atomic Structure Related to the Periodic Table

1. The elements are arranged in groups in the periodic chart of the elements. The groups are the vertical columns labeled IA, IIA, and so on. There is a relationship between the group an element belongs to and the charge on the ion that it forms. This relationship is illustrated in the table below.

Element	Group in Periodic Chart	Arrangement of Electrons	Symbol for Ion
Na	IA	2 8 1	Na^{1+}
Mg	IIA	2 8 2	Mg^{2+}
Al	IIIA	2 8 3	Al^{3+}

Elements in Group IA lose one electron, elements in Group IIA lose two electrons, and elements in Group IIIA lose three electrons to form ions.

Use the periodic chart to find the charge on these ions.

a. K _____   b. Ca _____   c. B _____

2. Elements in Groups VA, VIA, and VIIA gain electrons to form ions, as shown below.

Element	Group in Periodic Chart	Arrangement of Electrons	Symbol for Ion
N	VA	2 5	N^{3-}
O	VIA	2 6	O^{2-}
F	VIIA	2 7	F^{1-}

Use the periodic chart to find the charge on these ions.

a. P _____   b. S _____   c. Cl _____

3. Here is a summary of the relationship between the position of the element on the periodic chart and the charge on its ion.

Group on Chart	IA	IIA	IIIA	VA	VIA	VIIA
Charge on Ion	+1	+2	+3	−3	−2	−1

You will notice that nothing has been said about group IVA, containing elements that have four electrons in the outer energy level. Group IVA elements tend to neither gain nor lose electrons and hence usually do not form ions. We will deal with this group later.

---

6. a. 2 8 6
   b. 2 gained
   c. S^{2-}

7. a. 2 8 5
   b. 3 gained
   c. P^{3-}
   d. 2 8 7
   e. 1 gained
   f. Cl^{1-}

1. a. K^{1+}
   b. Ca^{2+}
   c. B^{3+}

2. a. P^{3-}
   b. S^{2-}
   c. Cl^{1-}

Those elements in the B groups of the periodic chart have the last *two* energy levels only partially filled with electrons. Consequently, we cannot use their position in the periodic chart to predict the charge on their ions. We will learn later that they have variable charges.

Example:

Element	Group in Periodic Chart	Arrangement of Electrons
Mn	VIIB	2 · 8 13 2

Both levels can lose electrons.

Tell what charge the ions in the following groups would have.

Group	Charge on Ion	Group	Charge on Ion
IIIA	a. _____	IIA	b. _____
VIIA	c. _____	VIA	d. _____
IA	e. _____	VA	f. _____

---

4. Answers to frame 3.

  a. $+3$ b. $+2$ c. -1 d. -2 e. $+1$ f. -3

---

5. Give the symbol and charge of the following.

  a. proton _____   b. neutron _____   c. electron _____

---

6. What is the atomic number of the following?

  a. Na _____   b. As _____   c. Br _____

5. a. p^{1+}
   b. n^0
   c. e^{-1}

---

7. Fill in the needed information.

Element	Number of Electrons	Arrangement of Electrons in Energy Levels 1 2 3 4
Mg	a. _____	b. _____
Cl	c. _____	d. _____
Fe	e. _____	f. _____
Cr	g. _____	h. _____

6. a. 11
   b. 33
   c. 35

---

8. Fill in the needed information.

Element	Arrangement in Energy Levels	Symbol and Charge of Ion
Ca	a. _____	b. _____
Al	c. _____	d. _____
F	e. _____	f. _____

7. a. 12
   b. 2 8 2
   c. 17
   d. 2 8 7
   e. 26
   f. 2 8 14 2
   g. 24
   h. 2 8 12 2

9. Complete the table.

Element	Arrangement in Energy Levels	Electron-Dot Symbol
S	a. _____	b. _____
N	c. _____	d. _____
Li	e. _____	f. _____

10. Complete this table.

Element	Group of the Periodic Chart	Symbol and Charge of Ion
Al	a. _____	b. _____
O	c. _____	d. _____
K	e. _____	f. _____

Part 7. Electron Arrangement in Energy Sublevels

1. The electron structure of the atom shown so far allows you to determine the electron arrangement in the main energy levels of an atom. These main energy levels have sublevels into which the electrons can be arranged. The number of sublevels contained in each main level is shown.

Main Level	Number of Sublevels
1	1
2	2
3	3
4	4
.	.
.	.
.	.

The first main level has 1 sublevel, the second main level has 2 sublevels, and so on.

How many sublevels do these main levels have?

a.   Third main level _____   b.   Fourth main level _____

2. The sublevels have the identifying letters of s, p, d, and f. To completely identify a sublevel, the number of the main level it is in and the letter of the sublevel is given.

Main Level	Sublevels
1	1s
2	2s, 2p
3	3s, 3p, 3d
4	4s, 4p, 4d, 4f
5	5s, 5p, 5d, 5f
6	6s, 6p, 6d
7	7s

8. a. 2 8 8 2
   b. Ca^{2+}
   c. 2 8 3
   d. Al^{3+}
   e. 2 7
   f. F^{1-}

9. a. 2 8 6
   b. ·S̈:
   c. 2 5
   d. :N̈·
   e. 2 1
   f. Li·

10. a. IIIA
    b. Al^{3+}
    c. VIA
    d. O^{2-}
    e. IA
    f. K^{1+}

1. a. 3
   b. 4

Main levels 5, 6, and 7 could have more sublevels listed, but no element has enough electrons to fill them, so they are never used.

a. List the sublevels of the fifth main level. _____

b. List the sublevels of the third main level. _____

---

3. The maximum number of electrons in each sublevel can now be shown.

Sublevel	Maximum number of electrons in sublevel
1s	2
2s	2
2p	6
3s	2
3p	6
3d	10
4s	2
4p	6
4d	10
4f	14

You can see that the maximum number of electrons in each type of sublevel is as given.

Type	Maximum Number of Electrons	Notations Used for Level and Number of Electrons
s	2	s^2
p	6	p^6
d	10	d^{10}
f	14	f^{14}

State the maximum number of electrons in each sublevel.

a. 4f _____   b. 1s _____   c. 4d _____   d. 2p _____

---

4. Determining the number of electrons in the energy sublevels is complex because of the "rule of eight."

Example: What is the electron arrangement of the Zn atom, which has 30 electrons?

Start placing electrons in the main levels and sublevels, keeping in mind the maximum number of electrons each sublevel can hold.

$$1s^2 \, 2s^2 \, 2p^6 \, 3s^2 \, 3p^6$$

At this point we must stop because the third main level is an outer level and it contains 8 electrons. To make the third main level an inner level, add another main level with a pair of electrons.

$$1s^2 \, 2s^2 \, 2p^6 \, 3s^2 \, 3p^6 \, 4s^2$$

We can now go back and complete the third main level.

$$1s^2 \, 2s^2 \, 2p^6 \, 3s^2 \, 3p^6 \, 4s^2 \, 3d^{10}$$

The thirty electrons have been used up.

2. a. 5s, 5p, 5d, 5f
   b. 3s, 3p, 3d

3. a. 14
   b. 2
   c. 10
   d. 6

Write the electron structure of the following, showing the arrangement in the main levels and sublevels.

a.   Sc – 21 electrons _____

b.   Fe – 26 electrons _____

5. Answers to frame 4.

  a. $1s^2\,2s^2\,2p^6\,3s^2\,3p^6\,4s^2\,3d^1$
  b. $1s^2\,2s^2\,2p^6\,3s^2\,3p^6\,4s^2\,3d^6$

6. The order for placing electrons in the main levels and sublevels can be arrived at using the following diagram.

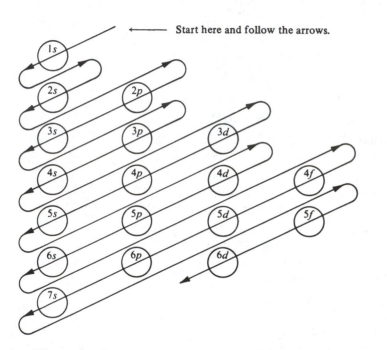

Start here and follow the arrows.

Example: What is the electron arrangement of the Rb atom, which has 37 electrons?

Follow the arrows in the diagram until 37 electrons are used up.

$$1s^2\,2s^2\,2p^6\,3s^2\,3p^6\,4s^2\,3d^{10}\,4p^6\,5s^1$$

Use the diagram to find the electron arrangement in the main levels and sublevels of these atoms.

a.   Na – 11 electrons _____

b.   Si   – 14 electrons _____

c.   Ti   – 22 electrons _____

d.   Ni   – 28 electrons _____

e.   Sn   – 50 electrons _____

f.   I   – 53 electrons _____

7. Answers to frame 6.

    a. $1s^2\,2s^2\,2p^6\,3s^1$

    b. $1s^2\,2s^2\,2p^6\,3s^2\,3p^2$

    c. $1s^2\,2s^2\,2p^6\,3s^2\,3p^6\,4s^2\,3d^2$

    d. $1s^2\,2s^2\,2p^6\,3s^2\,3p^6\,4s^2\,3d^8$

    e. $1s^2\,2s^2\,2p^6\,3s^2\,3p^6\,4s^2\,3d^{10}\,4p^6\,5s^2\,4d^{10}\,5p^2$

    f. $1s^2\,2s^2\,2p^6\,3s^2\,3p^6\,4s^2\,3d^{10}\,4p^6\,5s^2\,4d^{10}\,5p^5$

8. The electron arrangement of some elements will not follow the order in the diagram. Chromium and copper are examples of exceptions.

NAME _____

ATOMIC STRUCTURE
EVALUATION TEST 1
PART 2–PART 3 (frame 8)

1. Fill in the following chart.

Particle	Symbol	Charge
electron	a. _____	b. _____
neutron	c. _____	d. _____
e. _____	p	f. _____

2. Use the periodic chart to find the atomic number of the following.

a. Ca   b. I

3. Indicate which elements have the following atomic number.

a. 15   b. 79

4. Fill in the chart.

Element	Atomic Number	Number of Protons	Number of Electrons
Mg	a. _____	b. _____	c. _____
d. _____	20	e. _____	f. _____
g. _____	h. _____	14	i. _____
j. _____	k. _____	l. _____	8

5. What is the maximum number of electrons in the following energy levels?

a. second energy level   b. fourth energy level

1. a. _____
   b. _____
   c. _____
   d. _____
   e. _____
   f. _____
2. a. _____
   b. _____
3. a. _____
   b. _____
4. a. _____
   b. _____
   c. _____
   d. _____
   e. _____
   f. _____
   g. _____
   h. _____
   i. _____
   j. _____
   k. _____
   l. _____
5. a. _____
   b. _____

ATOMIC STRUCTURE
EVALUATION TEST 2
PART 3 (frame 9)–PART 5

1. Fill in the chart.

Element	Number of Electrons	Arrangement of Electrons in Energy Level 1 2 3 4
Li	a. _____	b. _____
N	c. _____	d. _____
Mg	e. _____	f. _____
S	g. _____	h. _____
Ca	i. _____	j. _____
Br	k. _____	l. _____
As	m. _____	n. _____

2. Fill in the chart.

Element	Arrangement in Levels 1 2 3 4	Electron-Dot Symbol
Na	a. _____	b. _____
Al	c. _____	d. _____
O	e. _____	f. _____
Cl	g. _____	h. _____
Ne	i. _____	j. _____

3. Fill in the chart.

Element	Arrangement in Levels 1 2 3 4	Number of Electrons Lost or Gained	Symbol and Charge of Ion
Li	a. _____	b. _____	c. _____
Mg	d. _____	e. _____	f. _____
Al	g. _____	h. _____	i. _____
N	j. _____	k. _____	l. _____
O	m. _____	n. _____	o. _____
F	p. _____	q. _____	r. _____

ATOMIC STRUCTURE
EVALUATION TEST 3
PART 6–PART 7

1. Fill in the chart.

Element	Arrangement in Levels	Group of Periodic Table
K	a. _____	b. _____
N	c. _____	d. _____
Ca	e. _____	f. _____
S	g. _____	h. _____
Cl	i. _____	j. _____

2. Fill in the chart.

Element	Group of Periodic Table	Electron-Dot Symbol
Na	a. _____	b. _____
O	c. _____	d. _____
Mg	e. _____	f. _____
Br	g. _____	h. _____
Al	i. _____	j. _____
N	k. _____	l. _____

3. Fill in the chart.

Element	Group of Periodic Table	Symbol and Charge of Ion
P	a. _____	b. _____
Na	c. _____	d. _____
Mg	e. _____	f. _____
S	g. _____	h. _____
F	i. _____	j. _____
B	k. _____	l. _____

4. Write the electron structure of the following elements showing the arrangement of the electrons in the main levels and sublevels. Use the diagram for assigning electrons.

a. Mg _____   d. Sr _____

b. S _____   e. I _____

c. V _____   f. Cs _____

UNIT 8
Chemical Bonding and Formula Writing

In this unit you will learn how compounds form and how ionic bonds and covalent bonds form compounds. You will learn the table of the most common ions and their oxidation numbers, and you will learn to write the formulas of simple compounds.

Part 1. Ionic and Covalent Bonding

1. We have learned that atoms will lose or gain electrons from the outermost energy level. When atoms lose or gain electrons, they form charged atoms or groups of atoms called *ions*. If the charges on these ions are opposite in sign, they will attract each other and form a chemical bond called an *ionic bond*. This bonding results in the formation of a compound.

   Example: Sodium has one electron in its outer level and chlorine has seven. If a sodium atom transfers its one outer electron to a chlorine atom, both atoms will have eight electrons in the outer energy level. In the process, they form oppositely charged ions that attract each other.

   Step 1. The electron is transferred.

   $$Na\odot \longrightarrow :\ddot{C}l:$$

   Step 2. Ions form. Sodium loses one electron and forms a + 1 ion. Chlorine gains one electron and forms a − 1 ion.

   $$Na \longrightarrow Na^{1+} \qquad Cl \longrightarrow Cl^{1-}$$

   Step 3. Ionic bonding takes place, resulting in the formation of a compound.

   $$Na^{1+} + Cl^{1-} \longrightarrow Na^{1+}Cl^{1-}$$

   └— Oppositely charged ions are attracted to each other.

   Ionic bonding is caused by the transfer of electrons from one atom to another.

   If ions are oppositely charged, they will _____ each other.

2. The chemical bond formed when oppositely charged ions are attracted to each other is called an _____.

   1. attract

3. Ionic bonding is caused by the _____ of electrons.

2. ionic bond

4. Elements in Group IVA of the periodic chart do not easily lose or gain electrons and hence form compounds in a different manner than ionic bonding. These elements *share* electrons to complete the outer energy level. This sharing of electrons forms a chemical bond called *covalent bonding*.

3. transfer

Example: Carbon has four electrons in its outer energy level and would like eight. Hydrogen has one electron in its outer energy level and would like two. If each of four hydrogen atoms shares its one electron with the same carbon atom, each hydrogen atom will have two electrons in its outer energy level and the carbon atom will have eight electrons in its outer energy level.

Step 1. Dots are used for the carbon-atom electrons and xs are used for the hydrogen-atom electrons.

	xH
:C·	xH
	xH
	xH
one carbon	four
atom	hydrogen
	atoms

Step 2. The electrons are shared.

$$
\begin{array}{c}
\text{H} \\
\text{x} \cdot \\
\text{Hx C xH} \\
\cdot \text{x} \\
\text{H}
\end{array}
$$

Step 3. Circles are drawn around each atom to illustrate the fact that as you count around each atom you will find two electrons around each hydrogen atom and eight electrons around the carbon atom.

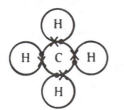

Covalent bonding is caused by the sharing of electrons between atoms.

When atoms have difficulty transferring electrons, they will _____ electrons.

5. The chemical bond formed by this sharing of electrons is called the _____ bond.

4. share

Part 2. Tables of Common Ions and Their Oxidation Number

1. You must learn the symbols and charges of the common ions before you can write the formulas of compounds. The charge of an ion is often called the *oxidation number*.

5. covalent

Another term for the charge of an ion is the _____.

2. The following is a table of the common positively charged ions. Except for silver, lead, and zinc, you need not memorize these. You can determine the charge or oxidation number of these ions by noting the element's position in the periodic chart.

1. oxidation number

THE COMMON POSITIVELY CHARGED IONS

Name of Ion	Symbol with Oxidation Number of Ion
hydrogen	H^{1+}
potassium	K^{1+}
silver	Ag^{1+}
sodium	Na^{1+}
barium	Ba^{2+}
calcium	Ca^{2+}
lead	Pb^{2+}
magnesium	Mg^{2+}
zinc	Zn^{2+}
aluminum	Al^{3+}

Give the symbol of the ion formed by the following elements. Remember that an ion must be written with the correct symbol and charge.

a. silver _____   b. zinc _____   c. aluminum _____

d. hydrogen _____   e. lead _____   f. barium _____

g. sodium _____   h. potassium _____   i. calcium _____

3. The following is a table of the common negatively charged ions. The oxidation number of these can be found by noting the position in the periodic chart. When these elements become negative ions, their names end in *ide*.

2. a. Ag^{1+}
   b. Zn^{2+}
   c. Al^{3+}
   d. H^{1+}
   e. Pb^{2+}
   f. Ba^{2+}
   g. Na^{1+}
   h. K^{1+}
   i. Ca^{2+}

THE COMMON NEGATIVELY CHARGED IONS

Name of Ion	Symbol with Oxidation Number of Ion
bromide	Br^{1-}
chloride	Cl^{1-}
fluoride	F^{1-}
iodide	I^{1-}
oxide	O^{2-}
sulfide	S^{2-}
nitride	N^{3-}
phosphide	P^{3-}

Give the symbol and charge of the ion.

a. fluoride _____   b. oxide _____   c. nitride _____

d. bromide _____   e. phosphide _____   f. sulfide _____

g. iodide _____   h. chloride _____

4. Give the symbol and charge of the ion formed by these elements.

   a. silver _____   b. chloride _____   c. sulfide _____

   d. potassium _____   e. zinc _____   f. iodide _____

   g. hydrogen _____   h. lead _____   i. nitride _____

   j. sodium _____   k. bromide _____   l. calcium _____

   m. magnesium _____   n. oxide _____   o. aluminum _____

3. a. F^{1-}
   b. O^{2-}
   c. N^{3-}
   d. Br^{1-}
   e. P^{3-}
   f. S^{2-}
   g. I^{1-}
   h. Cl^{1-}

5. Answers to frame 4.

   a. Ag^{1+} b. Cl^{1-} c. S^{2-} d. K^{1+} e. Zn^{2+} f. I^{1-} g. H^{1+} h. Pb^{2+}

   i. N^{3-} j. Na^{1+} k. Br^{1-} l. Ca^{2+} m. Mg^{2+} n. O^{2-} o. Al^{3+}

6. The ions we have looked at so far are monatomic, or one-atom, ions. There are also polyatomic, or many-atom, ions.

   Example: Na^{1+} —monatomic ion
   Cl^{1-} —monatomic ion
   $SO_4{}^{2-}$ —polyatomic ion
   $NH_4{}^{1+}$ —polyatomic ion

   The charge on the polyatomic ion pertains to the *entire* group of atoms forming the ion.

7. Which ions below are polyatomic ions? _____

   a. K^{1+} b. S^{2-} c. $NO_3{}^{1-}$ d. $CO_3{}^{2-}$

8. Which ions below are monatomic ions? _____

   a. $NH_4{}^{1+}$ b. H^{1+} c. N^{3-} d. $CO_3{}^{2-}$

7. c and d

9. The following is a table of the common polyatomic ions. The formula and charge of these polyatomic ions cannot be looked up on a chart. You will have to memorize them.

8. b and c

THE COMMON POLYATOMIC IONS

Name of Ion	Formula of Ion
ammonium	$NH_4{}^{1+}$
acetate	$C_2H_3O_2{}^{1-}$
hydroxide	OH^{1-}
nitrate	$NO_3{}^{1-}$
sulfate	$SO_4{}^{2-}$
carbonate	$CO_3{}^{2-}$
phosphate	$PO_4{}^{3-}$

Give the formula and charge of these polyatomic ions.

   a. nitrate _____   b. hydroxide _____

   c. acetate _____   d. ammonium _____

   e. carbonate _____   f. phosphate _____

10. Give the names of these polyatomic ions.

a. SO_4^{2-} _____

b. OH^{1-} _____

c. CO_3^{2-} _____

d. $C_2H_3O_2^{1-}$ _____

e. NH_4^{1+} _____

f. PO_4^{3-} _____

11. Give the formulas or symbols of these polyatomic and monatomic ions.

a. aluminum ion _____

b. phosphide ion _____

c. nitrate ion _____

d. iodide ion _____

e. acetate ion _____

f. hydroxide ion _____

g. lead ion _____

h. carbonate ion _____

i. nitride ion _____

j. ammonium ion _____

k. oxide ion _____

l. phosphate ion _____

Part 3. The Formula of a Compound

1. The formula of a compound is a shorthand method of expressing the chemical composition of a compound. We could write out the names of the monatomic ions and polyatomic ions that make up a compound, but this is cumbersome. Instead, we use the symbols for the monatomic ions and the formulas for the polyatomic ions.

Here are some examples of writing formulas.

Composition of Compound	Formula
sodium and chloride ions	NaCl
silver and nitrate ions	$AgNO_3$
ammonium and hydroxide ions	NH_4OH

Write the formulas for these compounds.

Composition of Compound	Formula
a. sodium ion and iodide ion	_____
b. potassium ion and bromide ion	_____
c. silver ion and chloride ion	_____
d. hydrogen ion and hydroxide ion	_____
e. ammonium ion and chloride ion	_____
f. ammonium ion and nitrate ion	_____

2. Here are some formulas of chemical compounds. Give their chemical compositions.

Formula	Composition of Compound
a. KF	_____
b. AgI	_____
c. HCl	_____
d. $NaC_2H_3O_2$	_____
e. KNO_3	_____
f. NH_4OH	_____

---

9. a. NO_3^{1-}
b. OH^{1-}
c. $C_2H_3O_2^{1-}$
d. NH_4^{1+}
e. CO_3^{2-}
f. PO_4^{3-}

10. a. sulfate
b. hydroxide
c. carbonate
d. acetate
e. ammonium
f. phosphate

11. a. Al^{3+}
b. P^{3-}
c. NO_3^{1-}
d. I^{1-}
e. $C_2H_3O_2^{1-}$
f. OH^{1-}
g. Pb^{2+}
h. CO_3^{2-}
i. N^{3-}
j. NH_4^{1+}
k. O^{2-}
l. PO_4^{3-}

1. a. NaI
b. KBr
c. AgCl
d. HOH
e. NH_4Cl
f. NH_4NO_3

3. Nothing has been said yet about the order in which we write symbols in a formula. The rule is that monatomic ions and polyatomic ions with positive charges are written first; monatomic ions and polyatomic ions with negative charges are written second. The square and circle below will help you to visualize this rule. Ions with a positive charge are written in the square and ions with a negative charge are written in the circle. Remember that the "square" always precedes the "circle."

In the compound KF, K would go in the square because it has a + charge. F would go in the circle because it has a − charge.

Below is a list of monatomic ions and polyatomic ions. Which would you put in the square? _____

a. Na^{1+} b. Ba^{2+} c. Cl^{1-} d. NH_4^{1+} e. NO_3^{1-}

2. a. potassium and fluoride ions
   b. silver and iodide ions
   c. hydrogen and chloride ions
   d. sodium and acetate ions
   e. potassium and nitrate ions
   f. ammonium and hydroxide ions

4. The ions Na^{1+}, Ba^{2+}, and NH_4^{1+} would be written first in a formula because they have a _____ charge.

3. a, b, and d

5. Here are more monatomic ions and polyatomic ions. Which would you put in the circle? _____

a. Pb^{2+} b. I^{1-} c. Zn^{2+} d. CO_3^{2-} e. OH^{1-}

4. positive

6. The ions I^{1-}, CO_3^{2-}, and OH^{1-} would be written last in a formula because they have a _____ charge.

5. b, d, and e

7. A compound must be neutral. This means that a compound cannot have a charge, even though the ions that compose it do. Another way of stating this is that there must be the same number of plus charges as there are minus charges in a compound. Below are illustrations of how we make the number of plus charges equal the number of minus charges in a compound.

6. negative

Compound: sodium and chloride ions

	Charge on Ion		Number of Ions Needed			Formula
Na	+1	X	1	= +1		$Na^{1+}Cl^{1-}$
Cl	−1	X	1	= −1		

The number of plus charges is equal to the number of minus charges; thus, the compound is neutral.

Here are several more examples.

Compound: barium and oxide ions

	Charge on Ion		Number of Ions Needed			Formula
Ba	+2	X	1	= +2		$Ba^{2+}O^{2-}$
O	−2	X	1	= −2		

The number of plus and minus charges is equal.

Compound: aluminum and nitride ions

	Charge on Ion		Number of Ions Needed			Formula
Al	+3	X	1	=	+3	$Al^{3+}N^{3-}$
N	−3	X	1	=	−3	

Again, the number of plus charges equals the number of minus charges.

Here are some examples for you to try. Fill in the blanks with the charge per ion and the number of ions needed.

Compound: calcium and sulfide ions

		Charge on Ion		Number of Ions Needed		
Ca	a. _____	X	b. _____	=	+2
S	c. _____	X	d. _____	=	−2

Compound: ammonium and nitrate ions

		Charge on Ion		Number of Ions Needed		
NH_4	e. _____	X	f. _____	=	+1
NO_3	g. _____	X	h. _____	=	−1

Compound: aluminum and phosphate ions

		Charge on Ion		Number of Ions Needed		
Al	i. _____	X	j. _____	=	+3
PO_4	k. _____	X	l. _____	=	−3

---

8. In the examples given in frame 7 the charges on each ion were the same. Here are some examples in which the charges are not the same.

Compound: calcium and chloride ions

	Charge on Ion		Number of Ions Needed			
Ca	+2	X	1	=	+2	(not equal)
Cl	−1	X	1	=	−1	

You can see that two chlorine ions are needed to provide a −2 charge. Thus, for a compound of calcium and chlorine, we have the following.

	Charge on Ion		Number of Ions Needed			
Ca	+2	X	1	=	+2	(equal)
Cl	−1	X	2	=	−2	

7. a. +2
   b. 1
   c. −2
   d. 1
   e. +1
   f. 1
   g. −1
   h. 1
   i. +3
   j. 1
   k. −3
   l. 1

Here is an example using polyatomic ions.

Compound: ammonium and sulfate ions

	Charge on Ion		Number of Ions Needed		
NH_4	+1	X	1	= +1	(not equal)
SO_4	−2	X	1	= −2	

Two NH_4 ions are needed to provide a +2 charge.

	Charge on Ion		Number of Ions Needed		
NH_4	+1	X	2	= +2	(equal)
SO_4	−2	X	1	= −2	

---

9. Fill in the blanks with the charge on the ion and the number of ions needed.

Compound: sodium and iodide ions

		Charge on Ion		Number of Ions Needed	
Na	a. _____	X	b. _____	= +1
I	c. _____	X	d. _____	= −1

Compound: calcium and fluoride ions

		Charge on Ion		Number of Ions Needed	
Ca	e. _____	X	f. _____	= +2
F	g. _____	X	h. _____	= −2

Compound: potassium and nitride ions

		Charge on Ion		Number of Ions Needed	
K	i. _____	X	j. _____	= +3
N	k. _____	X	l. _____	= −3

Compound: calcium and sulfate ions

		Charge on Ion		Number of Ions Needed	
Ca	m. _____	X	n. _____	= +2
SO_4	o. _____	X	p. _____	= −2

Compound: aluminum and nitrate ions

		Charge on Ion		Number of Ions Needed	
Al	q. _____	X	r. _____	= +3
NO_3	s. _____	X	t. _____	= −3

Compound: hydrogen and carbonate ions

		Charge on Ion		Number of Ions Needed	
H	u. _____	X	v. _____	= +2
CO_3	w. _____	X	x. _____	= −2

Compound: ammonium and phosphate ions

	Charge on Ion		Number of Ions Needed		
NH_4	y. _____	X	z. _____	=	+3
PO_4	aa. _____	X	bb. _____	=	−3

---

10. The number of ions of each type needed in a compound is expressed as a subscript in the formula. The subscript 1 is never written but is implied.

    Here are some examples.

    Compound: lead and iodide ions

	Charge on Ion		Number of Ions Needed		
Pb	+2	X	1	=	+2
I	−1	X	2	=	−2

    The number of ions is written below each element as a subscript in the formula. (Remember that the subscript 1 is never written.)

    Formula: PbI_2

    This says that one Pb ion combines with two I ions.

    Compound: hydrogen and chloride ions

	Charge on Ion		Number of Ions Needed		
H	+1	X	1	=	+1
Cl	−1	X	1	=	−1

    Formula: HCl

    One H ion combines with one Cl ion.

    Compound: potassium and oxide ions

K	+1	X	2	=	+2
O	−2	X	1	=	−2

    Formula: K_2O

    Two K ions combine with one O ion.

    In the formula Al_2O_3 there are (a) _____ Al ions and (b) _____ O ions.

    In the formula $CaCl_2$ there are (c) _____ Ca ions and (d) _____ Cl ions.

---

11. We can now use a streamlined method of writing a formula.

    Example: Write the formula for the compound formed from aluminum ions and oxide ions.

    Step 1. Write the ions in the proper order, indicating their charges.

    $Al^{3+}O^{2-}$

    Step 2. Use the numerical value of the charges to arrive at the subscripts by "crossing them over."

    $Al_2^{3+}O_3^{2-} = Al_2O_3$

    Note that the charge of oxygen becomes the subscript of aluminum and the charge of aluminum becomes the subscript of oxygen. The sign of the charge is not used with the subscript.

9. a. +1
   b. 1
   c. −1
   d. 1
   e. +2
   f. 1
   g. −1
   h. 2
   i. +1
   j. 3
   k. −3
   l. 1
   m. +2
   n. 1
   o. −2
   p. 1
   q. +3
   r. 1
   s. −1
   t. 3
   u. +1
   v. 2
   w. −2
   x. 1
   y. +1
   z. 3
   aa. −3
   bb. 1

10. a. 2
    b. 3
    c. 1
    d. 2

Example: Write the correct formula for the compound composed of zinc ions and nitrogen ions.

Step 1: $Zn^{2+}N^{3-}$

Step 2: $Zn_3^{2+}N_2^{3-}$ = Zn_3N_2

Write the correct formula for each of the following, using the two steps shown in the examples above.

calcium and phosphide ions   a. _____

                                b. _____

aluminum and sulfide ions   c. _____

                                d. _____

magnesium and nitride ions   e. _____

                                f. _____

---

12. Remember that when one of the charges in the formula is −1 or −1, you do not write subscript 1; it is always implied.

    11. a. $Ca^{2+}P^{3-}$
       b. Ca_3P_2
       c. $Al^{3+}S^{2-}$
       d. Al_2S_3
       e. $Mg^{2+}N^{3-}$
       f. Mg_3N_2

Example: Write the formula for the compound composed of sodium and sulfur ions.

Step 1: $Na^{1+}S^{2-}$

Step 2: $Na_2^{1+}S^{2-}$ = Na_2S

           └─ The subscript 1 is not written but is implied.

Write the correct formula for the following.

a.   potassium and oxide ions _____

b.   aluminum and iodide ions _____

c.   calcium and bromide ions _____

d.   hydrogen and sulfide ions _____

---

13. Write the correct formula for these compounds.

    12. a. K_2O
       b. AlI_3
       c. $CaBr_2$
       d. H_2S

a.   silver and oxide ions _____

b.   lead and nitride ions _____

c.   barium and fluoride ions _____

d.   zinc and phosphide ions _____

e.   aluminum and chloride ions _____

---

14. Because a polyatomic ion acts as a single unit, its subscript applies to the entire ion. To indicate this fact, we place the ion in parentheses. We do not use parentheses when the subscript is 1, nor do we write the subscript.

    13. a. Ag_2O
       b. Pb_3N_2
       c. BaF_2
       d. Zn_3P_2
       e. $AlCl_3$

Examples:

$Ca(OH)_2$ This formula says that the compound contains one calcium ion and two hydroxide ions.

$(NH_4)_2S$ This formula says that the compound contains two ammonium ions and one sulfide ion.

$(NH_4)_3PO_4$ This formula says that the compound contains three ammonium ions and one phosphate ion.

What do these formulas say?

$(NH_4)_2CO_3$ a.   Number of NH_4^{1+} ions _____

b.   Number of CO_3^{2-} ions _____

$Al(OH)_3$ c.   Number of Al^{3+} ions _____

d.   Number of OH^{1-} ions _____

---

15. Which of these formulas are written correctly? _____

   a. $BaOH_2$ b. $Al(C_2H_3O_2)_3$ c. $(NH_4)_2CO_3$ d. $ZnNO_{3\,2}$

14. a. 2   c. 1
    b. 1   d. 3

---

16. Writing the formulas of compounds that contain polyatomic ions can be done by using the charges for subscripts.

   Example: Write the formula of the compound composed of magnesium and the hydroxide ions.

   Step 1.   Write the ions in the proper order, indicating their charges.

     $Mg^{2+}OH^{1-}$

   Step 2.   Use the numerical value of the charges to arrive at the subscripts.

$$Mg^{2+}(OH)_2^{1-} = Mg(OH)_2$$

     The 1 is implied.   Parentheses are used because there is more than one poly-atomic ion.

   Example: Write the formula of the compound composed of calcium and phosphate ions.

   Step 1. $Ca^{2+}PO_4^{3-}$

   Step 2. $Ca_3^{2+}(PO_4)_2^{3-}$

   Write the formula with the correct subscripts for the following compounds.

   a. $CaNO_3$ _____   b. $AlOH$ _____

   c. NH_4S _____   d. NH_4OH _____

15. b and c

---

17. Given the names of the ions, write the formulas with the proper subscripts.

Composition of Compound	*Formula*
a. zinc and hydroxide ions	_____
b. aluminum and nitrate ions	_____
c. ammonium and fluoride ions	_____
d. ammonium and phosphate ions	_____

16. a. $Ca(NO_3)_2$
    b. $Al(OH)_3$
    c. $(NH_4)_2S$
    d. NH_4OH

---

18. Answers to frame 17.

   a. $Zn(OH)_2$ b. $Al(NO_3)_3$ c. NH_4F d. $(NH_4)_3PO_4$

---

19. The subscript 1 is never written but is _____ .

---

20. When a subscript is used with a polyatomic ion, _____ must be placed around the ion.

19. implied

21. Write the correct formula for the following compounds.

Composition of Compound	Formula
a. sodium and oxide ions	_____
b. potassium and nitrate ions	_____
c. aluminum and sulfate ions	_____
d. ammonium and phosphate ions	_____
e. zinc and iodide ions	_____
f. barium and acetate ions	_____
g. calcium and nitride ions	_____
h. aluminum and phosphate ions	_____

20. parentheses

22. Answers to frame 21.

a. Na_2O b. KNO_3 c. $Al_2(SO_4)_3$ d. $(NH_4)_3PO_4$
e. ZnI_2 f. $Ba(C_2H_3O_2)_2$ g. Ca_3N_2 h. $AlPO_4$

23. When the positive and negative charges in a compound are numerically equal, no subscripts are needed.

Example: $Ca^{2+}O^{2-}$ is *not* written Ca_2O_2. The lowest possible ratio is used when writing subscripts. The correct formula is CaO.

Which of the following formulas are written correctly? _____

a. Ca_2S_2 b. AlN c. $NaCl$ d. $Mg_2(SO_4)_2$

24. Write the formula for the compound composed of the following ions.

a. barium and oxide ions _____

b. zinc and sulfide ions _____

c. sodium and nitrate ions _____

d. aluminum and phosphate ions _____

23. b and c

25. The following statements can be made about formulas.

a. A chemical formula is a shorthand method of stating which elements make up a compound.
b. Positively charged ions are written first in a chemical compound and negative ions second.
c. A compound must be neutral (have no charge).
d. Subscripts are used to balance the charges in a compound.
e. The symbol for a polyatomic ion must be put in parentheses when a subscript is used with the ion.
f. When the charges in a compound are numerically equal, no subscripts are needed.
g. The subscript 1 is never written.

24. a. BaO
    b. ZnS
    c. $NaNO_3$
    d. $AlPO_4$

26. Write the formulas of the compounds composed of the following ions.

a. magnesium and chloride ions _____

b. aluminum and oxide ions _____

c. potassium and sulfide ions _____

d. hydrogen and fluoride ions _____

e. silver and sulfide ions _____

f. lead and iodide ions _____

g. sodium and oxide ions _____

h. hydrogen and bromide ions _____

i. calcium and carbonate ions _____

j. sodium and sulfate ions _____

k. aluminum and sulfide ions _____

l. magnesium and nitride ions _____

m. barium and nitrate ions _____

n. aluminum and hydroxide ions _____

o. ammonium and iodide ions _____

p. ammonium and sulfide ions _____

q. ammonium and acetate ions _____

r. ammonium and carbonate ions _____

s. ammonium and hydroxide ions _____

t. ammonium and phosphate ions _____

27. Answers to frame 26.

a. $MgCl_2$	b. Al_2O_3	c. K_2S	d. HF	e. Ag_2S
f. PbI_2	g. Na_2O	h. HBr	i. $CaCO_3$	j. Na_2SO_4
k. Al_2S_3	l. Mg_3N_2	m. $Ba(NO_3)_2$	n. $Al(OH)_3$	o. NH_4I
p. $(NH_4)_2S$	q. $NH_4C_2H_3O_2$	r. $(NH_4)_2CO_3$	s. NH_4OH	t. $(NH_4)_3PO_4$

Part 4. Writing Lewis Formulas

1. Lewis formulas (sometimes called electron dot structures) are written based on the number of electrons in the outer most energy levels of the atoms involved. When atoms form covalent bonds, they will share enough electrons so that each atom will have eight electrons in its outer energy level.

   Example: Two chlorine atoms can form a covalent bond. Each chlorine atom has seven electrons in its outer most energy level. To get eight electrons in its outer energy level, each chlorine atom will have to share one electron.

   Before forming covalent bond.

   Ċl   +   Ċl   2,8,7

   After forming covalent bond.

   Ċl:Ċl

   └── Shared pair of electrons

   As you count the electrons around each chlorine atom, you will find there are eight electrons around each atom. The shared pair in the middle is counted either way.

Draw the Lewis formula for the Br_2 molecule.

2. The oxygen molecule has a double covalent bond. Each oxygen atom has six electrons in its outer energy level and must share two electrons to get a configuration of eight electrons in its outer level.

Before forming covalent bond.   :Ö·   +   ·Ö:

After forming covalent bond.   :Ö::Ö:

⎡ Two shared pairs of electrons
⎣ or a double covalent bond.

Nitrogen has five electrons in its outer most energy level. It must share three electrons with another nitrogen atom. The N₂ molecule has a triple bond.

Draw the Lewis formula for N₂.

1. :B̈r:B̈r:

3. Hydrogen is an exception to the rule of eight. Hydrogen has only one energy level and it can only hold a maximum of two electrons. Two hydrogen atoms form a single covalent bond.

   Example: H : H

   Hydrogen can share electrons with other elements.

   Example: 4H · + ·Ċ· ⟶ H:C̈:H with H above and H below

   There are four single covalent bonds in CH_4.

   The steps in writing the above formula are:

   1. Write the electron dot formula of each atom.
   2. Determine how many electrons are needed by each atom to fill its outer energy level. In the above example, hydrogen needed one more and carbon needed four.
   3. Draw the Lewis formula.

Draw the Lewis formula for NH_3.

2. N̈ : : : N̈

4. Draw the Lewis formulas for the following.

   a. I_2

   b. H_2O

   c. HF

   d. PH_3

   e. CCl_4

3. H:N̈:H with H below

5. Answers to frame 4.

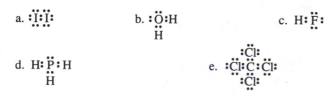

a. :Ï:Ï: b. :Ö:H with H below   c. H:F̈:

d. H:P̈:H with H below   e. :C̈l:C̈:C̈l: with :C̈l: above and :C̈l: below

NAME _____

CHEMICAL BONDING AND
FORMULA WRITING
EVALUATION TEST 1
PART 1–PART 2

1. Ionic bonding is caused by the _____ of electrons.

2. When electrons are shared, a _____ bond is formed.

3. Give the symbol and charge of the following ions.

    a. zinc   b. lead

    c. silver   d. magnesium

    e. potassium   f. aluminum

    g. barium   h. calcium

    i. hydrogen   j. sodium

4. Give the symbol and charge of the ion that the element forms.

    a. oxygen   b. sulfur

    c. fluorine   d. chlorine

    e. nitrogen   f. bromine

    g. iodine   h. phosphorus

5. Give the symbol and charge of the polyatomic ion.

    a. phosphate   b. ammonium

    c. hydroxide   d. carbonate

    e. acetate   f. sulfate

    g. nitrate

1. _____

2. _____

3. a. _____

   b. _____

   c. _____

   d. _____

   e. _____

   f. _____

   g. _____

   h. _____

   i. _____

   j. _____

4. a. _____

   b. _____

   c. _____

   d. _____

   e. _____

   f. _____

   g. _____

   h. _____

5. a. _____

   b. _____

   c. _____

   d. _____

   e. _____

   f. _____

   g. _____

CHEMICAL BONDING AND
FORMULA WRITING
EVALUATION TEST 2
PART 3 AND PART 4

Write the formulas for the following compounds.

1.   hydrogen and chloride ions   1. _____

2.   calcium and iodide ions   2. _____

3.   barium and oxide ions   3. _____

4.   aluminum and bromide ions   4. _____

5.   potassium and oxide ions   5. _____

6.   sodium and sulfide ions   6. _____

7.   aluminum and nitride ions   7. _____

8.   zinc and nitride ions   8. _____

9.   aluminum and sulfide ions   9. _____

10. hydrogen and sulfate ions   10. _____

11. zinc and nitrate ions   11. _____

12. aluminum and hydroxide ions   12. _____

13. calcium and hydroxide ions   13. _____

14. barium and phosphate ions   14. _____

15. magnesium and acetate ions   15. _____

16. ammonium and chloride ions   16. _____

17. aluminum and carbonate ions   17. _____

18. ammonium and nitrate ions   18. _____

19. ammonium and sulfide ions   19. _____

20. ammonium and phosphate ions   20. _____

21. ammonium and sulfate ions   21. _____

Draw the Lewis formula for these molecules.

22. F_2

23. HBr

UNIT 9
Naming Compounds

In this unit you will learn to identify and name the major classes of compounds: binary, ternary, and compounds composed of two polyatomic ions. You will learn to name ternary oxy-acids, ternary oxy-salts, and salts with more than one positive ion.

Part 1. The Major Classes of Compounds

1. The four major classes of compounds can be identified by noting what positive and negative ions compose these compounds. The first class of compounds is the *acids*.

   Examples: a. HCl —The positive ion is the monatomic hydrogen ion.
   The negative ion is the monatomic chloride ion.
   b. H_2S —The positive ion is the monatomic hydrogen ion.
   The negative ion is the monatomic sulfide ion.
   c. H_2SO_4 —The positive ion is the monatomic hydrogen ion.
   The negative ion is the sulfate ion.

   Note that in all of the examples above the positive monatomic ion is the hydrogen ion. Acids can be identified by determining whether the positive monatomic ion in the compound is the hydrogen ion. There are other acids that do not have the hydrogen ion in their formula. We will not deal with those here. You will learn about them later.

   Which of the following are acids? _____

   a. NaOH   b. NH_4Cl c. HBr   d. H_3PO_4

2. The compound below is missing a positive ion.

   $\boxed{?}\, Cl^{1-}$

   What is the positive ion if the compound is an acid? _____

   1. c and d

3. The second major class of compounds is the *bases*.

   Examples: a. NaOH   —The positive ion is the monatomic sodium ion.
   The negative ion is the hydroxide ion.
   b. $Ca(OH)_2$ —The positive ion is the monatomic calcium ion.
   The negative ion is the hydroxide ion.
   c. NH_4OH —The positive ion is the ammonium ion.
   The negative ion is the hydroxide ion.

   2. H^{1+}

Bases can be identified by the presence of the hydroxide ion as the negative component of the compound. Just as in the formula of acids, not all bases contain the hydroxide ion. In this unit we will only deal with bases that contain the hydroxide ion.

Which of the following are bases? _____

  a. CaO   b. $Al(OH)_3$ c. $Mg(OH)_2$ d. KOH

---

4. The following compound is missing the negative ion.

$$Na^{1+} \boxed{?}$$

What is the negative ion if the compound is a base? _____

3. b, c, and d

---

5. A third major class of compounds is the *oxides*.

   Examples: a. CaO —The positive ion is the monatomic calcium ion.
                          The negative ion is the monatomic oxide ion.
           b. Al_2O_3—The positive ion is the monatomic aluminum ion.
                          The negative ion is the monatomic oxide ion.

All oxides have the monatomic oxide ion as the negative component of the compound.

Which of the following are oxides? _____

  a. MgO   b. K_2O c. HI   d. NH_4I

4. OH^{1-}

③

---

6. Many ions contain oxygen atoms as part of a polyatomic ion. Compounds containing these ions are often mistaken for oxides. Look at the two lists below. Column A lists compounds that are made up of polyatomic ions that contain oxygen. Do not confuse these with the compounds in Column B., which are compounds composed of the monatomic oxide ion and are true oxides.

5. a and b

A	B
$CaCO_3$	K_2O
$NaNO_3$	MgO
$MgSO_4$	Al_2O_3
$AlPO_4$	NO
$KC_2H_3O_2$	CO_2
NaOH	SO_2

Which of these compounds are oxides? _____

  a. KOH   b. ZnO   c. $CaSO_4$ d. FeO

---

7. Which of these compounds are *not* oxides? _____

  a. CuO   b. SnO   c. $FeCO_3$ d. K_2SO_4

6. b and d

---

8. The following compound is missing the negative ion.

$$Mg^{2+} \boxed{?}$$

What is the negative ion if the compound is an oxide? _____

7. c and d

---

9. The fourth major class of compounds is the *salts*. If a compound is not identifiable as an acid, base, or oxide, it is probably a salt. In this unit, only the four major classes of compounds will be discussed. Therefore, you may assume that any compound not an acid, base, or oxide is definitely a salt. A salt may be composed of two monatomic ions, two polyatomic ions, or a monatomic ion and a polyatomic ion.

8. O^{2-} ion

④

Examples:

Which of the following compounds are salts? _____

a. CaO   b. HNO_3 c. NH_4I d. $Al(NO_3)_3$

---

10. Identify these compounds as acids, bases, oxides, or salts.

a. BaO _____

b. H_2SO_4 _____

c. NaI _____

d. $Ca(OH)_2$ _____

e. $CaSO_4$ _____

f. H_2S _____

g. $Fe(OH)_2$ _____

h. NH_4NO_3 _____

9. c and d

Part 2. Naming Binary Compounds

1. A binary compound is one that contains two, and *only* two, different elements.

   Examples: $NaCl$, $CaBr_2$, Al_2O_3

   Note that in a binary compound such as Al_2O_3 we have only two different elements, but we can have more than one atom of each type represented in the formula.

   Which of the compounds below are binaries? _____

   a. $CaCO_3$ b. HCl   c. NaOH   d. AlI_3

10. a. oxide
b. acid
c. salt
d. base
e. salt
f. acid
g. base
h. salt

2. Simple binaries are named by giving the name of the positive monatomic ion and adding "ide" to the first one or two syllables or stem of the name of the negative monatomic ion.

   Example:

   Name of binary compound: sodium chloride.

1. b and d

Here are more examples.

		Name of Positive Ion	+	Stem of Negative Ion, plus ide		
a.	KBr	= potassium	+	bromide	=	potassium bromide
b.	CaO	= calcium	+	oxide	=	calcium oxide
c.	MgS	= magnesium	+	sulfide	=	magnesium sulfide

Name these simple binary compounds.

		Name of Positive Ion	+	Stem of Negative Ion, plus ide		
NaF	=	_____	+	_____	= a.	_____
$AlCl_3$	=	_____	+	_____	= b.	_____
K_2S	=	_____	+	_____	= c.	_____
Ag_3N	=	_____	+	_____	= d.	_____

---

3. There are some elements that can have more than one oxidation number. We will look at three of these *variable* oxidation-number elements, Cu, Fe, and Sn, and learn how to name the binary compounds they form.

$$Cu^{1+} \text{ and } Cu^{2+}$$
$$Fe^{2+} \text{ and } Fe^{3+}$$
$$Sn^{2+} \text{ and } Sn^{4+}$$

Because these elements have variable oxidation numbers, they can form compounds in different ratios.

Example: $FeCl_2$ and $FeCl_3$

We cannot call both of these compounds simply "iron chloride" because we would have no way of differentiating the two compounds. We differentiate between them by using a roman numeral to indicate which oxidation number of iron exists in the compound.

Example: $Fe^{2+}Cl_2^{1-}$—Iron has an oxidation number of +2 in this compound; so we write the name as "iron (II) chloride."

Here are more examples.

a. $Fe^{3+}Cl_3^{1-}$ = iron (III) chloride
b. $Cu^{2+}O^{2-}$ = copper (II) oxide

The numerical value of the charge on the positive ion becomes the roman numeral in the name.

Name these binaries having variable oxidation numbers.

a. $Sn^{2+}F_2^{1-}$ _____

b. $Sn^{4+}F_4^{1-}$ _____

c. $Fe^{2+}O^{2-}$ _____

d. $Fe_2^{3+}O_3^{2-}$ _____

2. a. sodium fluoride
   b. aluminum chloride
   c. potassium sulfide
   d. silver nitride

4. You have already learned how to use charges to determine subscripts in a compound. The reverse process can be used when the subscripts are known and the charges are not. Cross the subscripts over to obtain the charges.

Example: Fe_2O_3 —To determine the charge on Fe, cross over the subscripts.

$Fe_2^{3+}O_3^{2-}$ —The charge on Fe is 3+.

Here are more examples.

$$CuCl_2 \longrightarrow Cu^{2+}Cl_2^{1-}$$

The 1 is implied.

$$SnI_4 \longrightarrow Sn^{4+}I_4^{1-}$$

What is the charge of the positive ion in the following compounds?

a. Fe_2S_3 _____   b. $SnBr_2$ _____

c. CuF_2 _____   d. $SnCl_4$ _____

5. If both <u>subscripts in</u> a binary formula are <u>1, a</u> different method must be used to determine the charge of the positive ion.

Example: CuO—Both subscripts are implied as 1. You learned that when both charges are numerically the same the subscripts are both 1. Therefore, when both subscripts are 1, the charges must be the same. If you know the charge of the negative ion, you know the charge of the positive ion. (The common negatively charged ions were listed on page 137 in frame 3.) In the CuO compound. The charge of the oxygen ion is −2 so the charge of the copper ion must be +2.

Here are more examples.

FeS—The charge of the sulfur ion is −2.
  The charge of the iron ion must be +2.
SnO—The charge of the ions must be $Sn^{2+}O^{2-}$.

What is the charge of the positive variable ion in these compounds?

a. CuS _____   b. FeO _____   c. SnS _____   d. CuCl _____

6. Name these binary compounds.

a. CuBr _____   b. FeS _____

c. CuO _____   d. SnO _____

7. Identify these compounds as simple or variable binaries and name them.

Compound	Simple or Variable Binary		Name
AlN	a. _____	b.	_____
K_2S	c. _____	d.	_____
CuO	e. _____	f.	_____
$FeCl_3$	g. _____	h.	_____
CaI_2	i. _____	j.	_____

3. a. tin (II) fluoride
   b. tin (IV) fluoride
   c. iron (II) oxide
   d. iron (III) oxide

4. a. Fe = +3
   b. Sn = +2
   c. Cu = +2
   d. Sn = +4

A

5. a. Cu = +2
   b. Fe = +2
   c. Sn = +2
   d. Cu = +1

6. a. copper (I) bromide
   b. iron (II) sulfide
   c. copper (II) oxide
   d. tin (II) oxide

8. When the positive hydrogen ion combines with one other element, it forms a <u>binary acid</u>.

Examples: HCl, H₂S, HBr

To name binary acids, use the word *hydro* for hydrogen and add *ic* to the stem of the negative monatomic ion. *The word* acid *must always be included.*

Example: HCl

hydrochloric acid

Here are more examples.

a.   HI = hydroiodic acid
b.   H₂S = hydrosulfuric acid

Name these binary acids.

a.   HF _____

b.   HCl _____

c.   HBr _____

9. Identify these compounds as simple, variable, or acid binaries and name them.

Compound	Type	Name
SnI₂	a. _____	b. _____
H₂S	c. _____	d. _____
MgO	e. _____	f. _____
HBr	g. _____	h. _____
CuCl	i. _____	j. _____
HCl	k. _____	l. _____

10. Certain ions that normally have a negative oxidation number can also have a variable positive oxidation number. Four elements in this category that we will consider are carbon, nitrogen, phosphorus, and sulfur.

Elements that normally have a negative oxidation number are called *nonmetals*. Compounds of these elements are called nonmetal compounds.

Examples: NO—Nonmetal nitrogen has a +2 oxidation number in this compound. Nonmetal oxygen has its normal −2 oxidation number.
CO—Nonmetal carbon has a +2 oxidation number in this compound. Nonmetal oxygen has its normal −2 oxidation number.

Compounds formed of two nonmetals are most easily identified by determining whether the positive component of the compound is carbon, nitrogen, phosphorus, or sulfur. In other compounds, these elements would normally be the negative component.

The four nonmetals that can form variable positive oxidation numbers are

(a) _____ , (b) _____ ,

(c) _____ , (d) _____ .

7. a. simple
b. aluminum nitride
c. simple
d. potassium sulfide
e. variable
f. copper (II) oxide
g. variable
h. iron (III) chloride
i. simple
j. calcium iodide

8. a. hydrofluoric acid
b. hydrochloric acid
c. hydrobromic acid

9. a. variable
b. tin (II) iodide
c. acid
d. hydrosulfuric acid
e. simple
f. magneisum oxide
g. acid
h. hydrobromic acid
i. variable
j. copper (I) chloride
k. acid
l. hydrochloric acid

11. The following compounds all contain nitrogen and oxygen, but in each compound nitrogen has a different positive oxidation number.

$$N_2O \quad N = +1$$
$$O = -2$$
$$NO \quad N = +2$$
$$O = -2$$
$$NO_2^{-2} \quad N = +4$$
$$O = -2$$

To name binary compounds containing elements with variable positive oxidation numbers, we use prefixes to tell how many of each type of atom are present. The prefixes are as follows.

mono = 1	tetra = 4
di = 2	penta = 5
tri = 3	hexa = 6

For example, in the compound S_2Cl_2 there are two sulfur and two chlorine atoms. Thus, the compound is called *disulfur dichloride*.

Here are more examples.

CO_2 = carbon dioxide (The prefix *mono* is usually omitted, just as the subscript 1 is omitted.)

N_2O_3 = dinitrogen trioxide

When the name of the element begins with a vowel, the vowel is usually dropped from the prefix.

Example: N_2O_5 = dinitrogen pentoxide

Name these nonmetal binaries.

a. SO_2 _____

b. N_2O _____

c. SO_3 _____

d. P_2O_5 _____

---

12. What numbers do these prefixes represent?

a.   penta = _____

b.   tetra = _____

c.   tri = _____

d.   hexa = _____

---

13. Name these compounds and identify them as simple, variable, acid, or nonmetal binaries.

Compound		Name		Type
FeI_2	a. _____		b. _____
H_2S	c. _____		d. _____
CO_2	e. _____		f. _____
AlP	g. _____		h. _____
SO_2	i. _____		j. _____
NO_2	k. _____		l. _____

10. a. carbon
    b. nitrogen
    c. phosphorus
    d. sulfur

11. a. sulfur dioxide
    b. dinitrogen oxide
    c. sulfur trioxide
    d. diphosphorus pentoxide

12. a. 5
    b. 4
    c. 3
    d. 6

14. There are some binary compounds that are known only by a common name. You must know two of these.

$$H_2O = water$$
$$NH_3 = ammonia$$

(The NH_3 molecule must not be confused with the NH_4^{1+} ion.)

Name these compounds or ions.

a. NH_3 _____

b. H_2O _____

c. NH_4^{1+} _____

15. Name these compounds and identify them as simple, variable, acid, nonmetal, or common binaries.

Compound		Name		Type
ZnO	a. _____		b. _____
CuCl	c. _____		d. _____
NH_3	e. _____		f. _____
HF	g. _____		h. _____
NO_2	i. _____		j. _____
H_2O	k. _____		l. _____

16. Answers to frame 15.

a. zinc oxide   b. simple   c. copper (I) chloride   d. variable
e. ammonia   f. common   g. hydrofluoric acid   h. acid
i. nitrogen dioxide   j. nonmetal   k. water   l. common

13. a. iron (II) iodide
    b. variable
    c. hydrosulfuric acid
    d. acid
    e. carbon dioxide
    f. nonmetal
    g. aluminum phosphide
    h. simple
    i. sulfur dioxide
    j. nonmetal
    k. nitrogen dioxide
    l. nonmetal

14. a. ammonia
    b. water
    c. ammonium ion

Part 3. Naming Ternary Compounds

1. A ternary compound is one that contains three, and *only* three, different elements. More than one atom of each type may be represented in the formula. Ternaries often contain polyatomic ions.

   Examples: $NaNO_3$, H_2SO_4, NH_4Cl

   Which of the compounds below are ternaries? _____

   a. Al_2O_3 b. NaOH   c. $CaCO_3$ d. $CaCl_2$

2. One type of ternary you will learn how to name is composed of a positive monatomic ion and a polyatomic ion.

Examples: $NaNO_3$, $CaCO_3$

To name this type of compound, simply name the positive monatomic ion and then
name the polyatomic ion.

Examples:

	Name of Positive Monatomic Ion		Name of Polyatomic Ion		Name of Compound
$NaNO_3$ =	sodium	+	nitrate	=	sodium nitrate
$CaCO_3$ =	calcium	+	carbonate	=	calcium carbonate

Name these ternary compounds.

Compound	Name of Positive Monatomic Ion		Name of Polyatomic Ion		Name of Compound
$Mg(OH)_2$ =	_____	+	_____	= a.	_____
K_2SO_4 =	_____	+	_____	= b.	_____
Na_2CO_3 =	_____	+	_____	= c.	_____
$AlPO_4$ =	_____	+	_____	= d.	_____

3. Variable-charge ions can also combine with polyatomic ions to form compounds. We
use roman numerals to differentiate between the different charges.

Examples: $CuCO_3$ = copper (II) carbonate
              $Fe(OH)_3$ = iron (III) hydroxide

Name these compounds.

a. $SnSO_4$ _____

b. $CuNO_3$ _____

c. $FePO_4$ _____

d. $Sn(OH)_4$ _____

4. Ternaries can also be composed of a polyatomic ion and a negative monatomic ion.
To name these ternaries, give the name of the polyatomic ion and add *ide* to the stem
of the negative monatomic ion.

Examples:

	Name of Polyatomic Ion		Add ide to Stem of Negative Ion		Name of Compound
NH_4Cl =	ammonium	+	chloride	=	ammonium chloride
$(NH_4)_2S$ =	ammonium	+	sulfide	=	ammonium sulfide

(Note that the only positively charged polyatomic ion you know is the ammonium
ion.)

Name these ternary compounds.

a. NH_4Br _____

b. NH_4I _____

1. b and c

2. a. magnesium hydroxide
   b. potassium sulfate
   c. sodium carbonate
   d. aluminum phosphate

3. a. tin (II) sulfate
   b. copper (I) nitrate
   c. iron (III) phosphate
   d. tin (IV) hydroxide

5. Name these ternary compounds.

a. $Al(OH)_3$ _____

b. $FeSO_4$ _____

c. NH_4Cl _____

d. $MgCO_3$ _____

e. $Sn(OH)_2$ _____

f. $CuNO_3$ _____

6. There are four common ternary acids you must know. They are:

$$H_2SO_4 = \text{sulfuric acid}$$
$$HNO_3 = \text{nitric acid}$$
$$HC_2H_3O_2 = \text{acetic acid}$$
$$H_3PO_4 = \text{phosphoric acid}$$

Do not confuse these ternary acids with the binary acids.

Example: *Ternary*   *Binary*

H_2SO_4 = sulfuric acid H_2S = hydrosulfuric acid

Name the acids listed below, stating whether they are binary or ternary.

	Binary or Ternary	*Name*
HCl	a. _____	b. _____
HNO_3	c. _____	d. _____
H_3PO_4	e. _____	f. _____
H_2SO_4	g. _____	h. _____
H_2S	i. _____	j. _____
$HC_2H_3O_2$	k. _____	l. _____

Part 4. Naming Compounds Composed of Two Polyatomic Ions

1. Compounds can be composed of two polyatomic ions. To name them, simply give the names of the positively charged ion and the negatively charged ion.

Example:

Name of Positive Ion	Name of Negative Ion	Name of Compound
NH_4OH = ammonium + hydroxide		= ammonium hydroxide

Name these compounds.

a. $(NH_4)_2CO_3$ _____

b. $NH_4C_2H_3O_2$ _____

c. NH_4NO_3 _____

4. a. ammonium bromide
   b. ammonium iodide

5. a. aluminum hydroxide
   b. iron (II) sulfate
   c. ammonium chloride
   d. magnesium carbonate
   e. tin (II) hydroxide
   f. copper (I) nitrate

6. a. binary
   b. hydrochloric acid
   c. ternary
   d. nitric acid
   e. ternary
   f. phosphoric acid
   g. ternary
   h. sulfuric acid
   i. binary
   j. hydrosulfuric acid
   k. ternary
   l. acetic acid

2. Name the following compounds.

   a. H_2SO_4 _____

   b. AgCl _____

   c. FeS _____

   d. NH_4OH _____

   e. H_2S _____

   f. NO_2 _____

   g. $Ca(OH)_2$ _____

   h. Ag_3PO_4 _____

   i. $(NH_4)_2S$ _____

   j. N_2O_5 _____

   k. $HC_2H_3O_2$ _____

   l. $CuCO_3$ _____

   m. HNO_3 _____

   n. NH_3 _____

   o. CO_2 _____

   p. $NH_4C_2H_3O_2$ _____

   q. AlN _____

   r. $SnBr_4$ _____

   s. P_2O_5 _____

   t. $Fe(NO_3)_3$ _____

   u. HCl _____

   v. H_3PO_4 _____

1. a. ammonium carbonate
   b. ammonium acetate
   c. ammonium nitrate

3. Answers to frame 2.

   a. sulfuric acid   b. silver chloride   c. iron (II) sulfide
   d. ammonium hydroxide   e. hydrosulfuric acid   f. nitrogen dioxide
   g. calcium hydroxide   h. silver phosphate   i. ammonium sulfide
   j. dinitrogen pentoxide   k. acetic acid   l. copper (II) carbonate
   m. nitric acid   n. ammonia   o. carbon dioxide
   p. ammonium acetate   q. aluminum nitride   r. tin (IV) bromide
   s. diphosphorus pentoxide t. iron (III) nitrate   u. hydrochloric acid
   v. phosphoric acid

Part 5. Naming Ternary Oxy-Acids in a Series

1. Ternary oxy-acids contain hydrogen, oxygen, and one other element. There can be a series of these acid compounds containing the same three elements in different ratios.

   Examples: H_2SO_3 $HClO$
                  H_2SO_4 $HClO_2$
                                  $HClO_3$
                                  $HClO_4$

   Which compounds are ternary oxy-acids?

   a. HNO_2 b. HI c. $H_2C_2O_4$ d. H_2S

---

2. To name the ternary oxy-acids, the oxidation number of the middle element must be known. The oxidation number of hydrogen is assumed to be $+1$, and the oxidation number of oxygen is assumed to be -2. All of the oxidation numbers in a compound must add up to zero. Multiply the oxidation number of each element by its subscript to get its total charge.

   1. a and c

   Example: HNO_3

	Oxidation Number		Subscript		Total
H	+1	X	1	=	+1
N	?	X	1	=	?
O	-2	X	3	=	-6
					0 Sum must equal zero.
			N must be +5.		

   Find the oxidation number of chlorine in the following acid.

   Example: $HClO_2$

	Oxidation Number	Subscript	Total
H			
Cl			
O			

---

3. Find the oxidation number of the middle elements in these ternary oxy-acids.

   a. H_3PO_3 _____   b. $HBrO_2$ _____

   c. $HClO_4$ _____   d. H_2CO_3 _____

   2. Oxidation number of chlorine = +3

4. To name ternary oxy-acids in which there are two different acids in the series, place the ending *ous* after the stem of the element, other than hydrogen or oxygen, with the lower oxidation number. Place the ending *ic* after the stem of the element, other than hydrogen or oxygen, with the higher oxidation number. Always add the word *acid*.

3. a.   P = +3
   b.   Br = +3
   c.   Cl = +7
   d.   C = +4

Example: Name H_2SO_3 and H_2SO_4

Step 1.   First determine the oxidation number of sulfur in each acid.

H_2SO_3 S = +4 (lower oxidation number)
H_2SO_4 S = +6 (higher oxidation number)

Step 2.   Apply the rules for naming.

Stem and
  Ending

H_2SO_3 – sulfurous acid
H_2SO_4 – sulfuric acid

Name these acids.

a. HNO_2 _____   b. HNO_3 _____

c. H_3PO_3 _____   d. H_3PO_4 _____

5. To name ternary oxy-acids when there are four acids in a series, use the prefix *hypo* before the stem of the element in the acid that has a lower oxidation number than the *ous* acid. Use the prefix *per* before the stem of the element in the acid that has a higher oxidation number than the *ic* acid. Add the word *acid* to the name.

4. a.   Nitrous acid
   b.   Nitric acid
   c.   Phosphorous acid
   d.   Phosphoric acid

Example: Name this series of acids.

   HClO, HClO₂, HClO₃, and HClO₄

Step 1.   First rank the acids, lowest first, according to the oxidation number of chlorine.

Formula	Oxidation Number of Cl
HClO	+1
HClO₂	+3
HClO₃	+5
HClO₄	+7

Step 2.   Name according to the rules.

Formula	Oxidation Number of Cl	Name
HClO	+1	Hypochlorous acid
HClO₂	+3	Chlorous acid
HClO₃	+5	Chloric acid
HClO₄	+7	Perchloric acid

Name this ternary oxy-acid series.
          HBrO, HBrO₂, HBrO₃, and HBrO₄

Formula	Oxidation Number of Br	Name

6. Answers to frame 5.

HBrO	+1	hypobromous acid
$HBrO_2$	+3	bromous acid
$HBrO_3$	+5	bromic acid
$HBrO_4$	+7	perbromic acid

Part 6. Naming Salts of Ternary Oxy-Acids

1. If the hydrogen element in a ternary oxy-acid is replaced with a metal, a ternary oxy-salt is obtained. The *ous* and *ic* endings of the acids are replaced with *ite* and *ate*. When needed, the prefixes *hypo* and *per* are used with the name of the metal.

Example: Name the salts Na_2SO_3 and Na_2SO_4.

Acid Derived From	Oxidation Number of S	Salt Formula	Name of Salt
H_2SO_3	+4	Na_2SO_3	Sodium sulfite
H_2SO_4	+6	Na_2SO_4	Sodium sulfate

Example: Name the salts $NaClO$, $NaClO_2$, $NaClO_3$, and $NaClO_4$.

Acid Derived From	Oxidation Number of Cl	Salt Formula	Name of Salt
HClO	+1	NaClO	Sodium hypochlorite
$HClO_2$	+3	$NaClO_2$	Sodium chlorite
$HClO_3$	+5	$NaClO_3$	Sodium chlorate
$HClO_4$	+7	$NaClO_4$	Sodium perchlorate

a. Name the ternary oxy-salts $NaNO_2$ and $NaNO_3$.

Acid Derived From	Oxidation Number of N	Salt Formula	Name of Salt

b. Name the ternary oxy-salts $KClO$, $KClO_2$, $KClO_3$, and $KClO_4$.

Acid Derived From	Oxidation Number of Cl	Salt Formula	Name of Salt

2. Answers to frame 1.

a.	HNO_2	+3	$NaNO_2$	sodium nitrite
	HNO_3	+5	$NaNO_3$	sodium nitrate
b.	HClO	+1	KClO	potassium hypochlorite
	$HClO_2$	+3	$KClO_2$	potassium chlorite
	$HClO_3$	+5	$KClO_3$	potassium chlorate
	$HClO_4$	+7	$KClO_4$	potassium perchlorate

Part 7. Naming Salts With More Than One Positive Ion

1. To name salts with more than one positive ion, name each positive ion and the negative ion.

   Examples: *Salt*   *Name*

   $NaHSO_4$ Sodium hydrogen sulfate

   $MgNH_4PO_4$ Magnesium ammonium phosphate

   Name the salts with more than one positive ion.

   Salt   *Name*

   $NaKSO_4$ a. _____

   $NaHCO_3$ b. _____

   $KAl(SO_4)_2$ c. _____

   $NaHS$ d. _____

---

2. Find the oxidation number of the middle elements in these compounds.

   a. HNO_2 _____   b. $HBrO_4$ _____

   c. HIO_3 _____   d. H_3PO_2 _____

1. a. Sodium potassium sulfate
   b. Sodium hydrogen carbonate
   c. Potassium aluminum sulfate
   d. Sodium hydrogen sulfide

---

3. Name these acids.

   a. H_2SO_3 _____

   b. H_2SO_4 _____

   c. $HClO$ _____

   d. $HClO_4$ _____

   e. $HClO_2$ _____

   f. $HClO_3$ _____

2. a. N = +3
   b. Br = +7
   c. I = +5
   d. P = +1

---

4. Name these ternary oxy-salts.

   a. $NaClO_4$ _____

   b. $KClO_3$ _____

   c. $NaClO$ _____

   d. $KClO_2$ _____

3. a. Sulfurous acid
   b. Sulfuric acid
   c. Hypochlorous acid
   d. Perchloric acid
   e. Chlorous acid
   f. Chloric acid

---

5. Name these salts with more than one positive ion.

   a. $NaHSO_4$ _____

   b. $Al(HCO_3)_3$ _____

   c. NH_4HS _____

   d. $Ca(HSO_4)_2$ _____

4. a. Sodium perchlorate
   b. Potassium chlorate
   c. Sodium hypochlorite
   d. Potassium chlorite

---

6. Answers to frame 5.

   a. Sodium hydrogen sulfate   b. Aluminum hydrogen carbonate
   c. Ammonium hydrogen sulfide   d. Calcium hydrogen sulfate

NAMING COMPOUNDS
EVALUATION TEST 1
PART 1–PART 2

State whether the following compounds are acids, bases, salts, or oxides.

1. $Fe(OH)_2$ 2. NH_4OH

3. NaF 4. HNO_3

5. K_2O 6. H_2S

Identify these compounds as simple, variable, acid, nonmetal, or common binaries and name them.

	Compound	Type	Name
7.	H_2O	a. _____	b. _____
8.	AlN	a. _____	b. _____
9.	Na_3P	a. _____	b. _____
10.	Fe_2O_3	a. _____	b. _____
11.	$CuCl$	a. _____	b. _____
12.	SnI_4	a. _____	b. _____
13.	HF	a. _____	b. _____
14.	H_2S	a. _____	b. _____
15.	HBr	a. _____	b. _____
16.	NO_2	a. _____	b. _____
17.	P_2O_5	a. _____	b. _____
18.	CO_2	a. _____	b. _____
19.	CCl_4	a. _____	b. _____
20.	NH_3	a. _____	b. _____
21.	CO	a. _____	b. _____

NAMING COMPOUNDS
EVALUATION TEST 2
PART 3–PART 4

Identify these compounds as binary or ternary acids and name them.

	Compound	Binary or Ternary Acid	Name
1.	HI	a. _____	b. _____
2.	H_2SO_4	a. _____	b. _____
3.	H_2S	a. _____	b. _____
4.	HBr	a. _____	b. _____
5.	HNO_3	a. _____	b. _____
6.	HCl	a. _____	b. _____
7.	$HC_2H_3O_2$	a. _____	b. _____
8.	H_3PO_4	a. _____	b. _____

Identify these compounds as composed of one positive monatomic ion and one polyatomic ion, one polyatomic ion and one negative monatomic ion, or two polyatomic
ions and name them.

	Compound	Type	Name
9.	$CaCO_3$	a. _____	b. _____
10.	$(NH_4)_2S$	a. _____	b. _____
11.	$(NH_4)_3PO_4$	a. _____	b. _____
12.	$CuSO_4$	a. _____	b. _____
13.	$Fe(OH)_3$	a. _____	b. _____
14.	NH_4Cl	a. _____	b. _____
15.	$NH_4C_2H_3O_2$	a. _____	b. _____
16.	$Sn(NO_3)_2$	a. _____	b. _____
17.	NH_4OH	a. _____	b. _____
18.	$(NH_4)_2SO_4$	a. _____	b. _____

NAMING COMPOUNDS
EVALUATION TEST 3
PART 5–PART 7

Identify these compounds as ternary oxy-acids, salts of ternary oxy-acids, or salts with more than one positive ion.

1. KHS
2. $NaClO_4$
3. HBrO
4. HNO_2
5. $CaSO_3$
6. H_2SO_4

Name these ternary oxy-acids.

7. H_2SO_4
8. H_2SO_3
9. HNO_2
10. HNO_3
11. $HClO_2$
12. HClO
13. $HClO_4$
14. $HClO_3$

Name these ternary oxy-salts.

15. $NaBrO_3$
16. $NaBrO_2$
17. NaBrO
18. $NaBrO_4$

Name these salts with more than one positive ion.

19. $KHCO_3$
20. $NaAl(SO_4)_2$
21. KHS
22. $CaNH_4PO_4$

1. _____
2. _____
3. _____
4. _____
5. _____
6. _____
7. _____
8. _____
9. _____
10. _____
11. _____
12. _____
13. _____
14. _____
15. _____
16. _____
17. _____
18. _____
19. _____
20. _____
21. _____
22. _____

UNIT 10
Balancing Chemical Equations

In this unit you will learn the parts of a chemical equation, what redox is, and how to assign oxidation numbers. You will learn how to balance chemical equations using both the inspection method and the redox method.

Part 1. The Parts of a Chemical Equation

1. A chemical equation tells what happens when particles of matter interact. The reacting particles are called *reactants* and the new substances formed are called *products*. Thus, a chemical equation describes a chemical reaction.

   Example:

$$Ca + S \longrightarrow CaS$$

   reactants   The arrow   product
   means "yields."

   In word form, this equation says that the reactants (Ca and S) react to yield a product (CaS).

   In the equation $Al + P \longrightarrow AlP$, the reactants are (a) _____ and (b) _____.

---

2. The product that is formed in the chemical equation $Mg + S \longrightarrow MgS$ is _____.

---

3. The arrow in a chemical equation means _____ .

---

4. A chemical equation describes a chemical _____ .

---

5. There are different types of chemical reactions. In one type there is only one reactant but there are two products.

   Example: $H_2O \longrightarrow H_2 + O_2$

   In the reaction, water is the reactant and hydrogen and oxygen are the products.

   In the equation $CaCO_3 \longrightarrow CaO + CO_2$, the reactant is (a) _____ and the products are (b) _____ and (c) _____ .

---

1. a. Al
   b. P

2. MgS

3. yields

4. reaction

6. The interacting particles need not be atoms; they can be molecules as well.

Example: NaOH + HCl $\longrightarrow$ NaCl + H$_2$O

The reactants are NaOH (sodium hydroxide) and HCl (hydrochloric acid). The products are NaCl (sodium chloride) and H$_2$O (water).

In the equation AgNO$_3$ + NaCl $\longrightarrow$ AgCl + NaNO$_3$, the reactants are (a) _____ and (b) _____. The products are (c) _____ and (d) _____ .

Part 2. Balancing Equations by Inspection

1. We use *coefficients* to balance chemical equations.

Example: Mg + O$_2$ $\longrightarrow$ 2MgO
                              ↑
                       coefficient

The coefficient 2 means "double everything in the formula." Therefore, the formula 2MgO means that there are two magnesium atoms and two oxygen atoms represented. Here are other examples of using coefficients and a method to help you determine the number of each kind of atom represented.

Example: 3Fe$_2$O$_3$

Element	Subscript	X	Coefficient	=	Number of Atoms Represented
Fe	2	X	3	=	6
O	3	X	3	=	9

How many atoms of each type are represented by the formula 4Al$_2$O$_3$?

Element	Subscript	X	Coefficient	=	Number of Atoms Represented
Al	a. _____	X	b. _____	=	c. _____
O	d. _____	X	e. _____	=	f. _____

2. The formula below has both a coefficient and a polyatomic ion with a subscript. You must add a step for the subscript on the polyatomic ion when determining the number of atoms. Note that the subscript on the ion does not apply to aluminum, which is not part of the polyatomic ion.

Example: 2Al(OH)$_3$

Element	Subscript on Atom	X	Subscript on Polyatomic Ion	X	Coefficient	=	Number of Atoms Represented
Al	1	X	–	X	2	=	2
O	1	X	3	X	2	=	6
H	1	X	3	X	2	=	6

How many atoms of each type are represented by the formula 3Al$_2$(SO$_4$)$_3$?

Element	Subscript on Atom	X	Subscript on Polyatomic Ion	X	Coefficient	=	Number of Atoms Represented
Al	a. _____	X	b. _____	X	c. _____	=	d. _____
S	e. _____	X	f. _____	X	g. _____	=	h. _____
O	i. _____	X	j. _____	X	k. _____	=	l. _____

Answer column (right margin):

5. a. CaCO$_3$
   b. CaO
   c. CO$_2$

6. a. AgNO$_3$
   b. NaCl
   c. AgCl
   d. NaNO$_3$

1. a. 2
   b. 4
   c. 8
   d. 3
   e. 4
   f. 12

3. Answers to frame 2.

a. 2	b. name	c. 3	d. 6	e. 1	f. 3
g. 3	h. 9	i. 4	j. 3	k. 3	l. 36

4. How many atoms of each type are represented in these formulas?

$2H_2SO_4$ a. H = _____   b. S = _____   c. O = _____

$3Ca_3(PO_4)_2$ d. Ca = _____   e. P = _____   f. O = _____

5. Look at the way the products oxygen and hydrogen are written in the following equation.

$$2H_2O \longrightarrow 2H_2 + O_2$$

Hydrogen and oxygen form pairs with atoms of their own kind. Elements that form these molecules of two atoms of the same element are called *diatomic*. The diatomic elements are hydrogen, oxygen, nitrogen, fluorine, chlorine, bromine, and iodine. The formulas for their diatomic molecules are as follows.

$$H_2 \quad O_2 \quad N_2 \quad F_2 \quad Cl_2 \quad Br_2 \quad I_2$$

Which of the following are diatomic elements? _____

a.   chlorine   b.   oxygen   c.   sodium   d.   bromine

6. In the following chemical equation, which is the diatomic molecule? _____

$$2HgO \longrightarrow 2Hg + O_2$$

7. List the molecular formulas of the seven diatomic elements.

_____   _____   _____

_____   _____   _____

8. Atoms of elements that form pairs with their own kind are called _____ elements.

9. Chemical equations must be balanced because the reactants and products must weigh the same. There must be the same number of atoms of each kind on each side of the arrow. We cannot use subscripts to balance equations, because then we would have incorrect formulas. The steps needed to balance an equation by the *inspection method* are as follows.

Step 1. Count the number of atoms of each type on each side of the arrow and use a coefficient to balance the equation.

Example: one magnesium atom

$$Mg + O_2 \longrightarrow MgO$$

two   one
oxygen   oxygen
atoms   atom

$$Mg + O_2 \longrightarrow 2MgO$$

coefficient

---

4. a. 4
   b. 2
   c. 8
   d. 9
   e. 6
   f. 24

5. a, b, and d

6. O_2

7. $H_2, O_2, N_2, F_2, Cl_2, Br_2, I_2$

8. diatomic

10. Step 2. Adding a coefficient on one side of the equation often causes an imbalance in the equation. Whenever you add a coefficient to one side of an equation, the entire equation must be rechecked to make sure that it balances.

Example:

one magnesium atom

two magnesium atoms

$$Mg + O_2 \longrightarrow 2MgO$$

two oxygen atoms

two oxygen atoms

$$2Mg + O_2 \longrightarrow 2MgO$$

Now that another coefficient has been added, the entire equation must be rechecked again. You will find that everything balances this time.

Balance the following equation by the inspection method.

$$Zn + O_2 \longrightarrow ZnO$$

Step 1.

Step 2.

---

11. Balance the following equations by the inspection method.

a. $H_2 + Cl_2 \longrightarrow HCl$

b. $Al + I_2 \longrightarrow AlI_3$

c. $H_2 + Br_2 \longrightarrow HBr$

d. $Ag_2O \longrightarrow Ag + O_2$

e. $KClO_4 \longrightarrow KCl + O_2$

f. $Mg + CO_2 \longrightarrow C + MgO$

g. $Cl_2 + KI \longrightarrow KCl + I_2$

h. $HCl + ZnS \longrightarrow ZnCl_2 + H_2S$

i. $SO_2 + O_2 \longrightarrow SO_3$

j. $Na + Cl_2 \longrightarrow NaCl$

k. $Ca + H_2O \longrightarrow H_2 + Ca(OH)_2$

l. $NaBr + H_2SO_4 \longrightarrow Na_2SO_4 + HBr$

m. $K + O_2 \longrightarrow K_2O$

n. $Cl_2 + NaBr \longrightarrow Br_2 + NaCl$

o. $N_2 + H_2 \longrightarrow NH_3$

p. $Zn + HCl \longrightarrow ZnCl_2 + H_2$

q. $Ca + O_2 \longrightarrow CaO$

r. $P + O_2 \longrightarrow P_2O_5$

s. $Mg + N_2 \longrightarrow Mg_3N_2$

t. $MgCl_2 + AgNo_3 \longrightarrow AgCl + Mg(NO_3)_2$

u. $Al + MnO_2 \longrightarrow Al_2O_3 + Mn$

v. $Na + H_2O \longrightarrow NaOH + H_2$

w. $PbO_2 \longrightarrow PbO + O_2$

x. $NO_2 + H_2O \longrightarrow HNO_3 + NO$

y. $Al + H_2SO_4 \longrightarrow Al_2(SO_4)_3 + H_2$

z. $C_2H_6 + O_2 \longrightarrow CO_2 + H_2O$

aa. $NH_3 + O_2 \longrightarrow NO + H_2O$

bb. $Al + C \longrightarrow Al_4C_3$

10. Step 1.
$$Zn + O_2 \longrightarrow 2ZnO$$
Step 2.
$$2Zn + O_2 \longrightarrow 2ZnO$$

12. Answers to frame 11.

a. $H_2 + Cl_2 \longrightarrow 2HCl$
b. $2Al + 3I_2 \longrightarrow 2AlI_3$
c. $H_2 + Br_2 \longrightarrow 2HBr$
d. $2Ag_2O \longrightarrow 4Ag + O_2$
e. $KClO_4 \longrightarrow KCl + 2O_2$
f. $2Mg + CO_2 \longrightarrow C + 2MgO$
g. $Cl_2 + 2KI \longrightarrow 2KCl + I_2$
h. $2HCl + ZnS \longrightarrow ZnCl_2 + H_2S$
i. $2SO_2 + O_2 \longrightarrow 2SO_3$
j. $2Na + Cl_2 \longrightarrow 2NaCl$
k. $Ca + 2H_2O \longrightarrow H_2 + Ca(OH)_2$
l. $2NaBr + H_2SO_4 \longrightarrow Na_2SO_4 + 2HBr$
m. $4K + O_2 \longrightarrow 2K_2O$
n. $Cl_2 + 2NaBr \longrightarrow Br_2 + 2NaCl$
o. $N_2 + 3H_2 \longrightarrow 2NH_3$
p. $Zn + 2HCl \longrightarrow ZnCl_2 + H_2$
q. $2Ca + O_2 \longrightarrow 2CaO$
r. $4P + 5O_2 \longrightarrow 2P_2O_5$
s. $3Mg + N_2 \longrightarrow Mg_3N_2$
t. $MgCl_2 + 2AgNO_3 \longrightarrow 2AgCl + Mg(NO_3)_2$
u. $4Al + 3MnO_2 \longrightarrow 2Al_2O_3 + 3Mn$
v. $2Na + 2H_2O \longrightarrow 2NaOH + H_2$
w. $2PbO_2 \longrightarrow 2PbO + O_2$
x. $3NO_2 + H_2O \longrightarrow 2HNO_3 + NO$
y. $2Al + 3H_2SO_4 \longrightarrow Al_2(SO_4)_3 + 3H_2$
z. $2C_2H_6 + 7O_2 \longrightarrow 4CO_2 + 6H_2O$
aa. $4NH_3 + 5O_2 \longrightarrow 4NO + 6H_2O$
bb. $4Al + 3C \longrightarrow Al_4C_3$

Part 3. Redox

1. The term *redox* refers to the process by which elements gain and lose electrons during a chemical reaction. The term is a combination of the words *reduction* and *oxidation*.

2. The word *redox* is a combination of the words _____ and oxidation.

3. When an element gains electrons, we say it has been *reduced*.

   Example: $S^0 + 2e^- \longrightarrow S^{2-}$

   Note that when electrons are gained they show up on the left side of the equation.

   A sulfur atom is neutral, so we give it a charge, or oxidation number, of zero. Sulfur has 16 protons and 16 electrons. When we add two or more electrons, we get the following.

   $$(16 \text{ protons}) + (18 \text{ electrons}) = S^{2-}$$

   Here is an example of an element being reduced. What is the charge on its ion? _____

   $$P^0 + 3e^- \longrightarrow P^?$$

2. reduction

4. When an element gains electrons, it has been _____ .

3. P^{3-}

5. When electrons are gained, they show up on the _____ side of the equation.

4. reduced

6. When an element loses electrons, we say it has been *oxidized*.

5. left

   Example: $N^{3-} \longrightarrow N^0 + 3e^-$

   └─ Note that when electrons are
   lost they show up on the right
   side of the equation.

   A nitrogen ion has seven protons and ten electrons. When we take three electrons away, we get the following.

$$(+7 \text{ protons}) + (-7 \text{ electrons}) = N^0$$

   Here is an example of an ion being oxidized. What was the original charge on the ion? _____

$$S \longrightarrow S^0 + 2e^-$$

7. When an element loses electrons, it has been _____ .

6. S^{2-}

8. When electrons are lost, they show up on the _____ side of the equation.

7. oxidized

9. In which of the examples below has reduction taken place? _____

   a. $Cu^{2+} + 2e^- \longrightarrow Cu^0$

   b. $Cl^{1-} \longrightarrow Cl^0 + 1e^-$

8. right

10. In which of the examples below does oxidation take place? _____

   a. $Zn^{2+} + 2e^- \longrightarrow Zn^0$

   b. $Cu^{1+} \longrightarrow Cu^{2+} + 1e^-$

9. a

11. If you have trouble deciding whether an element has lost or gained electrons, use the following number line.

10. b

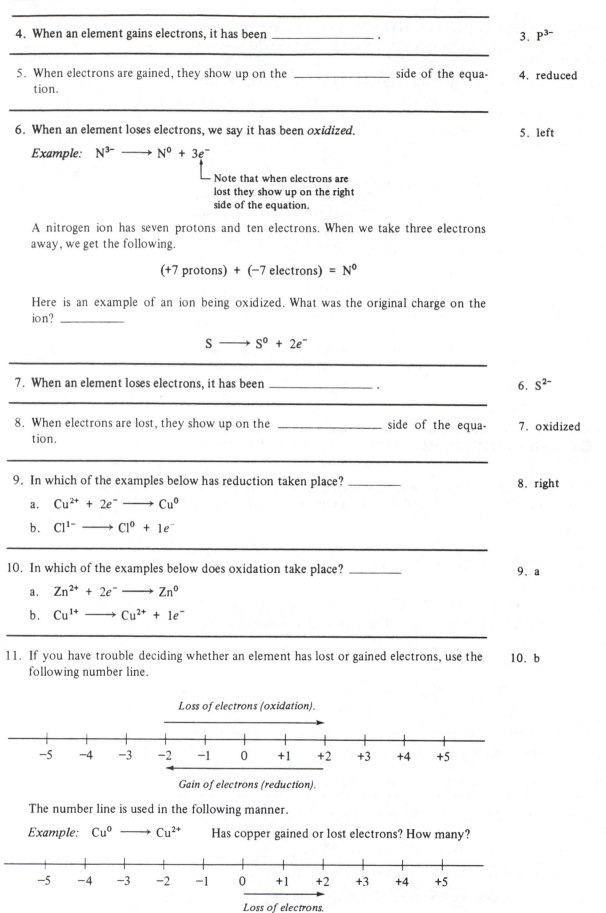

Loss of electrons (oxidation).

Gain of electrons (reduction).

The number line is used in the following manner.

Example: $Cu^0 \longrightarrow Cu^{2+}$ Has copper gained or lost electrons? How many?

Loss of electrons.

When an atom loses electrons, it becomes more positive. Copper started with a charge of zero and ended with a charge of plus two. The arrow is two units long, so copper lost two electrons.

Example: $N^{5+} \longrightarrow N^{3-}$ Has nitrogen gained or lost electrons? How many?

Gain of electrons.

When an atom gains electrons, it becomes more negative. Nitrogen started with a charge of +5 and ended with a charge of −3. The arrow is eight units long, so nitrogen gained eight electrons.

Find the gain and loss of electrons in the following, using the number line.

	Are electrons gained or lost?	*How many electrons are gained or lost?*
$Fe^{3+} \longrightarrow Fe^{2+}$	a. _____	b. _____
$Sn^{2+} \longrightarrow Sn^{4+}$	c. _____	d. _____
$Cl^{0} \longrightarrow Cl^{1-}$	e. _____	f. _____
$Cu^{2+} \longrightarrow Cu^{1+}$	g. _____	h. _____
$S^{2-} \longrightarrow S^{0}$	i. _____	j. _____

Part 4. Oxidation Numbers

1. The oxidation number of an element can vary, depending on what other elements it has combined with to form a compound. For this reason we need rules for determining oxidation numbers.

   RULE 1. All free elements (those elements not in a compound or polyatomic ion) have an oxidation number of zero.

   Example: Cu Cl_2 Na

   None of these elements has combined with any other element, so they are free elements. In the examples below, the elements are combined in compounds, so they are not free elements.

   $CuCl_2$ NaCl

   Which of the following are free elements? _____

   a. FeO   b. K   c. S   d. CO

2. Which of the examples below are *not* free elements? _____

   a. H_2S b. P   c. Al   d. NaOH

3. In the equation below, Fe and Cl_2 are free elements and are assigned oxidation numbers of zero.

   $$Fe^{0} + Cl_2{}^{0} \longrightarrow FeCl_3$$

   Assign a zero oxidation number to the free elements in this equation. _____

   $$CuO + NH_3 \longrightarrow N_2 + H_2O + Cu$$

11. a. gained
    b. 1
    c. lost
    d. 2
    e. gained
    f. 1
    g. gained
    h. 1
    i. lost
    j. 2

1. b and c

2. a and d

4. In the last equation in frame 3, which is the diatomic molecule? _____

3. $N_2{}^0$, Cu^0

5. RULE 2. Monatomic (one-atom) ions have an oxidation number that is the same as their charge. Some have a variable charge and hence a variable oxidation number.

4. $N_2{}^0$

Examples: Na^{1+} Cl^{1-}

In which examples below are the ions one-atom ions? _____

a. $SO_4{}^{2-}$ b. I^{1-} c. $NO_3{}^{1-}$ d. Mg^{2+}

6. Indicate the oxidation numbers of the following monatomic ions.

5. b and d

a. calcium _____   b. oxygen _____

c. sulfur _____   d. aluminum _____

7. In the equations we will be balancing, we will frequently find monatomic ions forming compounds.

6. a. +2
b. −2
c. −2
d. +3

Example: NaCl is made up of the Na^{1+} ion and the Cl^{1-} ion.

What monatomic ions make up the following compounds?

a. FeO _____   b. MgS _____

8. Which of the following compounds are made up only of monatomic ions? _____

7. a. Fe^{2+} and O^{2-}
b. Mg^{2+} and S^{2-}

a. NaOH   b. H_2SO_4 c. Fe_2O_3 d. NaBr

9. Assign oxidation numbers to the monatomic ions that make up the following compounds.

8. c and d

a. CuS _____   b. Na_2O _____

10. In the following equation, identify the compounds that are made up only of monatomic ions and assign oxidation numbers to these ions.

9. a. Cu^{2+} and S^{2-}
b. Na^{1+} and O^{2-}

$$Ca(OH)_2 \longrightarrow CaO + H_2O$$

a. Compounds composed only of monatomic ions _____

b. Oxidation numbers of these ions _____

11. Assign oxidation numbers to the monatomic ions that make up the compound in this equation.

10. a. CaO and H_2O
b. Ca^{2+} O^{2-} $H_2{}^{1+}$ O^{2-}

$$Fe + O_2 \longrightarrow Fe_2O_3$$

12. RULE 3. For practical purposes, we can assume that hydrogen will always have an oxidation number of +1 and oxygen will always have an oxidation number of −2 when they are combined with other elements in a compound.

Example: $H_2^0 + O_2^0 \longrightarrow H_2^{1+}O^{2-}$

Note that H_2 and O_2 are free elements, so their oxidation number is zero. In H_2O, the hydrogen and oxygen are combined in a compound, so they have oxidation numbers of +1 and −2.

Assign oxidation numbers to oxygen in the equation below. _____

$$2Ag_2O \longrightarrow 4Ag + O_2$$

13. Assign oxidation numbers to hydrogen in the equation below. _____

$$H_2 + Br_2 \longrightarrow 2HBr$$

14. RULE 4. All of the oxidation numbers in a compound must add up to zero.

Example: $CaCl_2$

	Oxidation Number		Number of Atoms		Total
Ca =	+2	X	1	=	+2
Cl =	−1	X	2	=	−2
					0 ← Sum must equal zero.

Try this one: Fe_2O_3.

	Oxidation Number		Number of Atoms		Total
Fe =	a. _____	X	b. _____	=	c. _____
O =	d. _____	X	e. _____	=	f. _____
					g. _____

15. Rules 3 and 4 can be used to determine the oxidation numbers of the elements in polyatomic ions.

Example: HNO_3 We assume that hydrogen is +1 and oxygen is −2. All oxidation numbers must add up to zero.

	Oxidation Number		Number of Atoms		Total
H =	+1	X	1	=	+1
N =	?	X	1	=	?
O =	−2	X	3	=	−6
					0 ← Sum must equal zero.

N must be +5.

What are the oxidation numbers of the elements in the following compounds?

a. H_2SO_4 _____   b. H_3PO_4 _____

---

11. $Fe_2^{3+}O_3^{2-}$

12. $Ag_2^{1+}O^{2-}$ and O_2^0

13. H_2^0 and $H^{1+}Br^{1-}$

14. a. +3
b. 2
c. +6
d. −2
e. 3
f. −6
g. 0

16. What are the oxidation numbers of the elements in this compound?

KIO$_3$ _____

17. The compound ammonia is an unusual compound in that it is an exception to the rule that the positive component is always written first in a formula.

Example:　$N^{3-}H_3^{1+}$

If we wrote the formula as H_3N, the compound would appear to be an acid, which it is not. When assigning oxidation numbers, remember that the positive and negative components are reversed in writing the formula for ammonia.

When writing the formula for ammonia, the negative oxidation number is written

(a) _____ and the positive oxidation number is written (b) _____ .

Part 5. Balancing Equations by the Redox Method

1. During a chemical reaction, elements often change their oxidation numbers. We can use this fact to help us balance equations. The first step in balancing an equation, using oxidation-number change, is to assign oxidation numbers to all of the elements—free and combined—in the chemical equation.

Step 1.　Assign oxidation numbers.

Example:　$Sn^0 + H^{1+}N^{5+}O_3^{2-} \longrightarrow Sn^{4+}O_2^{2-} + N^{4+}O_2^{2-} + H_2^{1+}O^{2-}$

Assign oxidation numbers to the elements in this equation. _____

$$Fe + O_2 \longrightarrow Fe_2O_3$$

2. Step 2.　Next locate the elements that have lost or gained electrons and connect with lines those elements on the left and right sides of the equation.

Example:　$Sn^0 + H^{1+}N^{5+}O_3^{2-} \longrightarrow Sn^{4+}O_2^{2-} + N^{4+}O_2^{2-} + H_2^{1+}O^{2-}$

Assign oxidation numbers to the elements in the following equation and draw lines to indicate the elements that have lost or gained electrons.

$$H_2 + Br_2 \longrightarrow HBr$$

15. a. H^{1+}, S^{6+}, O^{2-}
　　b. H^{1+}, P^{5+}, O^{2-}

16. K^{1+}, I^{5+}, O^{2-}

17. a. first
　　b. second

1. $Fe^0 + O_2^0 \longrightarrow Fe_2^{3+}O_3^{2-}$

3. Step 3. You must now determine how many electrons have been lost or gained by each element taking part in an electron transfer.

Example: $Sn^0 + H^{1+}N^{5+}O_3^{2-} \longrightarrow Sn^{4+}O_2^{2-} + N^{4+}O_2^{2-} + H_2^{1+}O^{2-}$

Element	Electrons gained or lost?	How many?
Sn	lost	4
N	gained	1

If you have trouble determining this, use a number line.

In the equation below, determine how many electrons have been lost or gained.

$$Mg^0 + S^0 \longrightarrow Mg^{2+}S^{2-}$$

Element	Electrons gained or lost?	How many?
Mg	a. _____	b. _____
S	c. _____	d. _____

4. Step 4. The number of electrons lost and the number of electrons gained must be equal. We use coefficients to accomplish this.

Example: $Sn^0 + H^{1+}N^{5+}O_3^{2-} \longrightarrow Sn^{4+}O_2^{2-} + N^{4+}O_2^{2-} + H_2^{1+}O^{2-}$

Coefficient = 1. Four electrons lost.

Coefficient = 4. One electron gained.

These coefficients are *not* automatically the coefficients used in the final equation. They merely tell the number of atoms of each type needed. In word form, the diagram above states that we will need four nitrogen atoms, each gaining one electron, to balance one tin atom, which loses four electrons.

What will the coefficients be for the equation below?

$Na^{1+}N^{5+}O_3^{2-} + Fe^0 \longrightarrow Na^{1+}N^{3+}O_2^{2-} + Fe_2^{3+}O_3^{2-}$

Coefficient = ?
Two electrons gained.

Coefficient = ?
Three electrons lost.

a. The coefficient for N should be _____.

b. The coefficient for Fe should be _____ .

5. Here is another equation. Find the coefficients that balance the electron loss and gain.

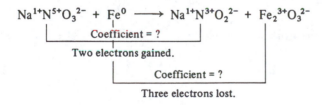

$Cu + HNO_3 \longrightarrow Cu(NO_3)_2 + NO + H_2O$

Coefficient = ?
Two electrons lost.

Coefficient = ?
Three electrons gained.

a. The coefficient for Cu should be _____ .

b. The coefficient for N should be _____ .

Answers (right margin):

2. $H_2^0 + Br_2^0 \longrightarrow H^{1+}Br^{1-}$

3. a. lost
   b. 2
   c. gained
   d. 2

4. a. 3
   b. 2

6. We can now use the four steps we have learned to balance a chemical equation using the redox method.

5. a. 3
   b. 2

Example: $Sn + HNO_3 \longrightarrow SnO_2 + NO_2 + H_2O$

Step 1. Assign oxidation numbers.

$$Sn^0 + H^{1+}N^{5+}O_3^{2-} \longrightarrow Sn^{4+}O_2^{2-} + N^{4+}O_2^{2-} + H_2^{1+}O^{2-}$$

Step 2. Locate the elements that have lost or gained electrons and connect them with lines.

$$Sn^0 + H^{1+}N^{5+}O_3^{2-} \longrightarrow Sn^{4+}O_2^{2-} + N^{4+}O_2^{2-} + H_2^{1+}O^{2-}$$

Step 3. Determine the number of electrons lost or gained by each element.

$$Sn^0 + H^{1+}N^{5+}O_3^{2-} \longrightarrow Sn^{4+}O_2^{2-} + N^{4+}O_2^{2-} + H_2^{1+}O^{2-}$$

Four electrons lost.

One electron gained.

Step 4. Use coefficients to balance the electron loss and gain.

$$Sn^0 + H^{1+}N^{5+}O_3^{2-} \longrightarrow Sn^{4+}O_2^{2-} + N^{4+}O_2^{2-} + H_2^{1+}O^{2-}$$

Coefficient = 1.

Four electrons lost.

Coefficient = 4.

One electron gained.

With the addition of two more steps we can now complete balancing the equation.

Step 5. The coefficients obtained in balancing the electron loss and gain represent the *number of atoms of each type* needed on each side of the equation.

$$Sn^0 + H^{1+}N^{5+}O_3^{2-} \longrightarrow Sn^{4+}O_2^{2-} + N^{4+}O_2^{2-} + H_2^{1+}O^{2-}$$

Coefficient = 1.

Four electrons lost.

Coefficient = 4.

One electron gained.

The coefficient of 1 indicates that one atom of Sn is needed on each side of the equation. The coefficient of 4 indicates that four atoms of N are needed on each side of the equation.

$$Sn + 4HNO_3 \longrightarrow SnO_2 + 4NO_2 + H_2O$$

Step 6. Balance the rest of the equation by inspection. Leave hydrogen and oxygen until last.

$$Sn + 4HNO_3 \longrightarrow SnO_2 + 4NO_2 + 2H_2O$$

This is a balanced equation.

7. Balance the following equation using the six steps you have learned.

$$NaNO_3 + Fe \longrightarrow NaNO_2 + Fe_2O_3$$

Step 1. Assign oxidation numbers.

Step 2. Locate elements that have lost or gained electrons.

Step 3. Determine the number of electrons lost or gained.

Step 4. Use coefficients to balance electrons lost or gained.

Step 5. Use coefficients to determine the number of atoms of each type needed on each side of the equation.

Step 6. Balance the rest of the equation by inspection, leaving hydrogen and oxygen for last.

8. Answers to frame 7.

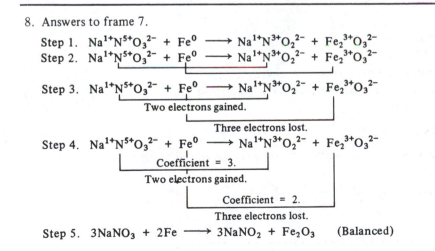

Step 1. $Na^{1+}N^{5+}O_3^{2-} + Fe^0 \longrightarrow Na^{1+}N^{3+}O_2^{2-} + Fe_2^{3+}O_3^{2-}$
Step 2. $Na^{1+}N^{5+}O_3^{2-} + Fe^0 \longrightarrow Na^{1+}N^{3+}O_2^{2-} + Fe_2^{3+}O_3^{2-}$

Step 3. $Na^{1+}N^{5+}O_3^{2-} + Fe^0 \longrightarrow Na^{1+}N^{3+}O_2^{2-} + Fe_2^{3+}O_3^{2-}$
              Two electrons gained.
                    Three electrons lost.
Step 4. $Na^{1+}N^{5+}O_3^{2-} + Fe^0 \longrightarrow Na^{1+}N^{3+}O_2^{2-} + Fe_2^{3+}O_3^{2-}$
              Coefficient = 3.
          Two electrons gained.
                    Coefficient = 2.
                    Three electrons lost.
Step 5. $3NaNO_3 + 2Fe \longrightarrow 3NaNO_2 + Fe_2O_3$ (Balanced)

9. Diatomic elements sometimes present a problem when balancing equations, because only even numbers of atoms can be present in a diatomic molecule.

Example: $Fe + O_2 \longrightarrow Fe_2O_3$

Step 1. $Fe^0 + O_2^0 \longrightarrow Fe_2^{3+}O_3^{2-}$

Step 2. $Fe^0 + O_2^0 \longrightarrow Fe_2^{3+}O_3^{2-}$

Step 3. $Fe^0 + O_2^0 \longrightarrow Fe_2^{3+}O_3^{2-}$

Three electrons lost.

Two electrons gained.

Step 4. $Fe^0 + O_2^0 \longrightarrow Fe_2^{3+}O_3^{2-}$

Coefficient = 2.

Three electrons lost.

Coefficient = 3.

Two electrons gained.

Step 5.   The coefficient of 3 tells us that we need three atoms of oxygen. Because oxygen comes in pairs (O_2), we can have only even numbers of oxygen atoms. We double *both* coefficients to avoid this problem.

$$4Fe^0 + 3O_2^0 \longrightarrow 2Fe_2^{3+}O_3^{2-}$$

Coefficient = 4.

Three electrons lost.

Coefficient = 6.

Two electrons gained.

Step 6.   This step is not needed for this equation.

Balance the following equation.

$$Fe + Cl_2 \longrightarrow FeCl_3$$

Step 1.

Step 2.

Step 3.

Step 4.

Step 5.

Step 6.

10. Answers to frame 9.

Step 1. $Fe^0 + Cl_2^0 \longrightarrow Fe^{3+}Cl_3^{1-}$

Step 2. $Fe^0 + Cl_2^0 \longrightarrow Fe^{3+}Cl_3^{1-}$

Step 3. $Fe^0 + Cl_2^0 \longrightarrow Fe^{3+}Cl_3^{1-}$

Three electrons lost.

One electron gained.

Step 4. $Fe + Cl_2 \longrightarrow FeCl_3$

Coefficient = 1.

Three electrons lost.

Coefficient = 3.

One electron gained.

(Double both coefficients)

Step 5. $2Fe + 3Cl_2 \longrightarrow 2FeCl_3$

Coefficient = 2.

Three electrons lost.

Coefficient = 6.

One electron gained.

Step 6. Not needed.

11. Balance these equations by the redox method.

a. $Fe + O_2 \longrightarrow Fe_2O_3$

b. $KClO_3 \longrightarrow KCl + O_2$

c. $H_2O + NaClO_3 + SO_2 \longrightarrow H_2SO_4 + NaCl$

d. $Na + H_2O \longrightarrow NaOH + H_2$

e. $H_2S + HNO_3 \longrightarrow S + NO + H_2O$

f. $Al + HCl \longrightarrow AlCl_3 + H_2$

g. $N_2 + H_2 \longrightarrow NH_3$

h. $Al + O_2 \longrightarrow Al_2O_3$

i. $NaClO_3 \longrightarrow NaCl + O_2$

j. $K + H_2O \longrightarrow KOH + H_2$

12. Answers to frame 11.

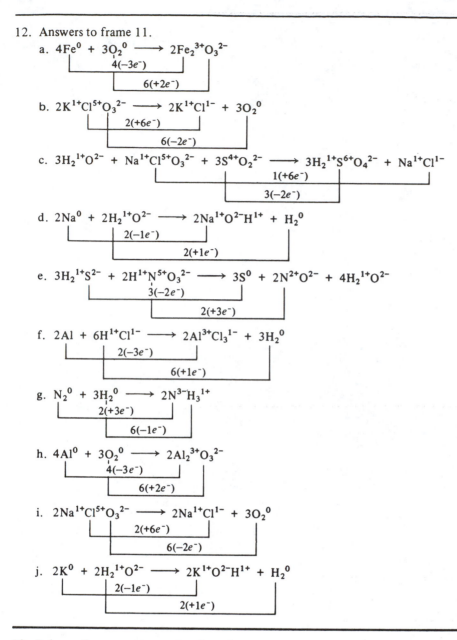

a. $4Fe^0 + 3O_2^0 \longrightarrow 2Fe_2^{3+}O_3^{2-}$
$4(-3e^-)$
$6(+2e^-)$

b. $2K^{1+}Cl^{5+}O_3^{2-} \longrightarrow 2K^{1+}Cl^{1-} + 3O_2^0$
$2(+6e^-)$
$6(-2e^-)$

c. $3H_2^{1+}O^{2-} + Na^{1+}Cl^{5+}O_3^{2-} + 3S^{4+}O_2^{2-} \longrightarrow 3H_2^{1+}S^{6+}O_4^{2-} + Na^{1+}Cl^{1-}$
$1(+6e^-)$
$3(-2e^-)$

d. $2Na^0 + 2H_2^{1+}O^{2-} \longrightarrow 2Na^{1+}O^{2-}H^{1+} + H_2^0$
$2(-1e^-)$
$2(+1e^-)$

e. $3H_2^{1+}S^{2-} + 2H^{1+}N^{5+}O_3^{2-} \longrightarrow 3S^0 + 2N^{2+}O^{2-} + 4H_2^{1+}O^{2-}$
$3(-2e^-)$
$2(+3e^-)$

f. $2Al + 6H^{1+}Cl^{1-} \longrightarrow 2Al^{3+}Cl_3^{1-} + 3H_2^0$
$2(-3e^-)$
$6(+1e^-)$

g. $N_2^0 + 3H_2^0 \longrightarrow 2N^{3-}H_3^{1+}$
$2(+3e^-)$
$6(-1e^-)$

h. $4Al^0 + 3O_2^0 \longrightarrow 2Al_2^{3+}O_3^{2-}$
$4(-3e^-)$
$6(+2e^-)$

i. $2Na^{1+}Cl^{5+}O_3^{2-} \longrightarrow 2Na^{1+}Cl^{1-} + 3O_2^0$
$2(+6e^-)$
$6(-2e^-)$

j. $2K^0 + 2H_2^{1+}O^{2-} \longrightarrow 2K^{1+}O^{2-}H^{1+} + H_2^0$
$2(-1e^-)$
$2(+1e^-)$

13. Balance these equations using the redox method.

a. $P + Br_2 \longrightarrow PBr_3$

b. $Mg + N_2 \longrightarrow Mg_3N_2$

c. $I_2 + Cl_2 + H_2O \longrightarrow HIO_3 + HCl$

d. $S + HNO_2 \longrightarrow H_2SO_3 + N_2O$

e. $F_2 + H_2O \longrightarrow HF + O_2$

f. $As_2O_3 + Cl_2 + H_2O \longrightarrow H_3AsO_4 + HCl$

g. $Cl_2 + KI \longrightarrow KCl + I_2$

h. $Na_2SO_3 + H_2O + I_2 \longrightarrow NaI + H_2SO_4$

i. $I_2 + H_2S \longrightarrow S + HI$

j. $Fe_2O_3 + CO \longrightarrow Fe + CO_2$

14. Answers to frame 13.

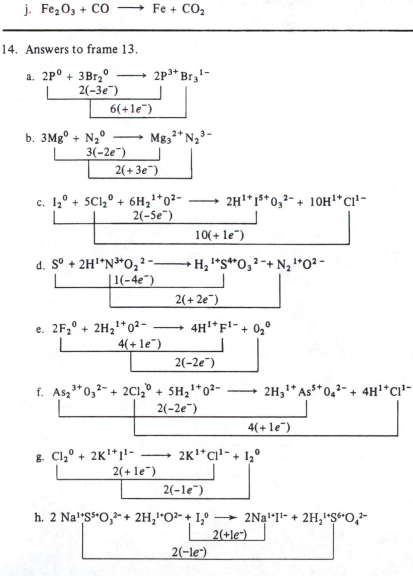

a. $2P^0 + 3Br_2^0 \longrightarrow 2P^{3+}Br_3^{1-}$
$2(-3e^-)$
$6(+1e^-)$

b. $3Mg^0 + N_2^0 \longrightarrow Mg_3^{2+}N_2^{3-}$
$3(-2e^-)$
$2(+3e^-)$

c. $I_2^0 + 5Cl_2^0 + 6H_2^{1+}O^{2-} \longrightarrow 2H^{1+}I^{5+}O_3^{2-} + 10H^{1+}Cl^{1-}$
$2(-5e^-)$
$10(+1e^-)$

d. $S^0 + 2H^{1+}N^{3+}O_2^{2-} \longrightarrow H_2^{1+}S^{4+}O_3^{2-} + N_2^{1+}O^{2-}$
$1(-4e^-)$
$2(+2e^-)$

e. $2F_2^0 + 2H_2^{1+}O^{2-} \longrightarrow 4H^{1+}F^{1-} + O_2^0$
$4(+1e^-)$
$2(-2e^-)$

f. $As_2^{3+}O_3^{2-} + 2Cl_2^0 + 5H_2^{1+}O^{2-} \longrightarrow 2H_3^{1+}As^{5+}O_4^{2-} + 4H^{1+}Cl^{1-}$
$2(-2e^-)$
$4(+1e^-)$

g. $Cl_2^0 + 2K^{1+}I^{1-} \longrightarrow 2K^{1+}Cl^{1-} + I_2^0$
$2(+1e^-)$
$2(-1e^-)$

h. $2 Na^{1+}S^{5+}O_3^{2-} + 2H_2^{1+}O^{2-} + I_2^0 \longrightarrow 2Na^{1+}I^{1-} + 2H_2^{1+}S^{6+}O_4^{2-}$
$2(+1e^-)$
$2(-1e^-)$

i. $I_2^0 + H_2^{1+}S^{2-} \longrightarrow S^0 + 2H^{1+}I^{1-}$

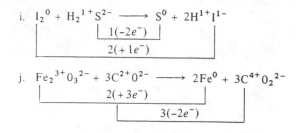

$$1(-2e^-)$$
$$2(+1e^-)$$

j. $Fe_2^{3+}O_3^{2-} + 3C^{2+}O^{2-} \longrightarrow 2Fe^0 + 3C^{4+}O_2^{2-}$

$$2(+3e^-)$$
$$3(-2e^-)$$

Part 6. Balancing Redox Equations Written in Ionic Form

1. Many compounds exist as ions in solution. These compounds are written not as formulas but as ions in an equation.

Example:

Molecular form: $HCl + NaOH \longrightarrow NaCl + H_2O$

Ionic form: $H^+ + Cl^- + Na^+ + OH^- \longrightarrow Na^+ + Cl^- + H_2O$

Notice that H_2O is not in ionic form. It is common for one or more compounds in a solution not to exist in ionic form.

Balance ionic redox equations using the same steps as in balancing non-ionic redox equations.

Example: Balance the following equation.

$$Fe^{2+} + Cl_2 \longrightarrow Fe^{3+} + Cl^{1-}$$

Step 1. Assign oxidation numbers. The monatomic ions have the same oxidation numbers as their charges. Cl_2 is a free element.

$$Fe^{2+} + Cl_2^0 \longrightarrow Fe^{3+} + Cl^{1-}$$

Step 2. Locate the elements that have lost or gained electrons.

$$Fe^{2+} + Cl_2^0 \longrightarrow Fe^{3+} + Cl^{1-}$$

Step 3. Determine how many electrons have been lost or gained by each element.

$$Fe^{2+} + Cl_2^0 \longrightarrow Fe^{3+} + Cl^{1-}$$

(1 lost)

(1 gained)

Step 4. Use coefficients to balance the electron loss or gain.

$$Fe^{2+} + Cl_2^0 \longrightarrow Fe^{3+} + Cl^{1-}$$

2 (1 lost)

2 (1 gained)

Step 5. Use the coefficients to balance the equation.

$$2Fe^{2+} + Cl_2^0 \longrightarrow 2Fe^{3+} + 2Cl^{1-}$$

Step 6. Balance the equation by inspection. Both the number of atoms and the total charge on each side of an ionic equation must balance. Multiply the charge on each ion by its coefficient, and add up the charges on each side.

$$2Fe^{2+} + Cl_2^0 \longrightarrow 2Fe^{3+} + 2Cl^{1-}$$

$$+4 \quad + \quad 0 = +4 \quad +6 \quad + \quad -2 = +4$$

Balance this ionic redox equation.

$$Sn^{2+} + Br_2 \longrightarrow Sn^{4+} + Br^{1-}$$

2. Answer to frame 1.

$$Sn^{2+} + Br_2^{\ 0} \longrightarrow Sn^{4+} + 2Br^{1-}$$

1 (2 lost)

2 (1 gained)

3. You must be careful to keep the oxidation numbers of the individual elements in a polyatomic ion separate from the total charge on a polyatomic ion.

Example: Find the oxidation number of the elements in the ion $HCO_3^{\ 1-}$.

Overall the charge on the ion is -1. The oxidation numbers on the individual elements must equal -1.

	Oxidation Number		Subscript	Charge
H	+1	X	1	+1
C	?	X	1	?
O	-2	X	3	-6
				-1 (overall charge on ion)

The oxidation number of carbon must equal +4.

Find the oxidation number of each element in the polyatomic ions.

a. NO_3^- b. $SO_4^{\ 2-}$ c. MnO_4^-

4. Answers to frame 3.

a. N = +5, O = -2
b. S = +6, O = -2
c. Mn = +7, O = -2

5. More-complex equations can now be balanced.

Example: Balance the ionic equation.

$$H^+ + Cu + SO_4^{\ 2-} \longrightarrow Cu^{2+} + SO_2 + H_2O$$

Step 1. $H^+ + Cu^0 + S^{6+}O_4^{\ 2-} \longrightarrow Cu^{2+} + S^{4+}O_2^{\ 2-} + H_2^{\ 1+}O^{2-}$

Step 2. $H^+ + Cu^0 + S^{6+}O_4^{\ 2-} \longrightarrow Cu^{2+} + S^{4+}O_2^{\ 2-} + H_2^{\ 1+}O^{2-}$

Step 3. $H^+ + Cu^0 + S^{6+}O_4^{\ 2-} \longrightarrow Cu^{2+} + S^{4+}O_2^{\ 2-} + H_2^{\ 1+}O^{2-}$

2 lost

2 gained

Step 4. $H^+ + Cu^0 + S^{6+}O_4^{\ 2-} \longrightarrow Cu^{2+} + S^{4+}O_2^{\ 2-} + H_2^{\ 1+}O^{2-}$

1 (2 lost)

1 (2 gained)

Step 5. $H^+ + Cu^0 + SO_4^{\ 2-} \longrightarrow Cu^{2+} + SO_2 + H_2O$

Step 6.   Balance charges and atoms. Balance the charges first.

Left Side	Right Side
1 H^+ = +1	1 Cu^{2+} = +2
1 $SO_4^{\ 2-}$ = -2	
$\overline{-1}$	

The charge on the left side does not equal the charge on the right side. Increase the H^+ ions to balance the charge.

Left Side	*Right Side*
$4\ H^+\ \ = +4$	$1\ Cu^{2+} = +2$
$1\ SO_4{}^{2-} = \underline{-2}$	
$\phantom{1\ SO_4{}^{2-} = }+2$	

The charge on each side is balanced using H^+ or OH^- ions. Complete the balancing of the equation by inspection.

$$4H^+ + Cu^0 + SO_4{}^{2-} \longrightarrow Cu^{2+} + SO_2 + 2H_2O$$

Balance these equations.

a. $MnO_4{}^- + S^{2-} + H^+ \longrightarrow Mn^{2+} + S + H_2O$

b. $FeS + NO_3{}^- + H^+ \longrightarrow NO + SO_4{}^{2-} + Fe^{3+} + H_2O$

c. $CuS + H^+ + NO_3{}^- \longrightarrow Cu^{2+} + S + NO + H_2O$

---

6. Answers to frame 5.

a. $2MnO_4{}^- + 5S^{2-} + 16H^+ \longrightarrow 2Mn^{+2} + 5S + 8H_2O$
b. $FeS + 3NO_3{}^- + 4H^+ \longrightarrow 3NO + SO_4{}^{2-} + Fe^{3+} + 2H_2O$
c. $3CuS + 8H^+ + 2NO_3{}^- \longrightarrow 3Cu^{2+} + 3S + 2NO + 4H_2O$

Part 7. Types of Chemical Reactions

1. There are four main types of chemical reactions. The first type is called a combination or synthesis reaction. In this unit the term combination will be used. Combination reactions have the general form:

$$A + B \longrightarrow AB$$

Notice that the equation has two reactants but only one product. In a combination reaction, two or more substances react to form a more complex substance.

The reactants in a combination reaction can be any combination of elements or compounds. The product will always be a compound.

Example: Two elements combine to form a compound.

$$2K + I_2 \longrightarrow 2KI$$

What will the product be in the following reaction?

$$Ca + S \longrightarrow$$

---

2. The reactants can be an element and a compound.

Example: $2CO + O_2 \longrightarrow 2CO_2$

What will the product be for this reaction?

$$2SO_2 + O_2 \longrightarrow$$

1. CaS

3. The reactants can also be two compounds.

Example: $H_2O + CO_2 \longrightarrow H_2CO_3$

What will the product be for this reaction?

$$H_2O + SO_2 \longrightarrow$$

4. State if the following combination reactions are between two elements, an element and a compound or two compounds.

a. $CuSO_4 + 5H_2O \longrightarrow CuSO_4 \cdot 5H_2O$ _____

b. $2Na + Cl_2 \longrightarrow 2NaCl$ _____

c. $2NO + O_2 \longrightarrow 2NO_2$ _____

5. A decomposition reaction is the reverse of a combination reaction. The reaction has the general form:

$$AB \longrightarrow A + B$$

Notice that the equation has one reactant but two products.

The reactant will always be a compound but the products can be any combination of elements and simple compounds.

Example: A compound decomposes to form two elements.

$$2H_2O \longrightarrow 2H_2 + O_2$$

What will the products be for this reaction?

$$2H_2O \longrightarrow$$

6. The products of a decomposition reaction can be an element and a simple compound.

Example: $2KClO_3 \longrightarrow 2KCl + O_2$

What will the products be for this reaction?

$$2NaClO_3 \longrightarrow$$

7. The products can also be two simpler compounds. The complex compound shown as a reactant is a hydrated compound. The dot in the formula of the compound indicates that the five water molecules are bonded to the copper (II) sulfate. This hydrated bond is broken during the reaction.

Example: $CuSO_4 \cdot 5H_2O \longrightarrow CuSO_4 + 5H_2O$

What will the products be for this reaction?

$$CaCl_2 \cdot 2H_2O \longrightarrow$$

8. State if the following decomposition reactions have two elements, an element and a compound or two compounds as products.

a. $2NaNO_3 \longrightarrow 2NaNO_2 + O_2$ _____

b. $2HCl \longrightarrow H_2 + Cl_2$ _____

c. $CaSO_4 \cdot H_2O \longrightarrow CaSO_4 + H_2O$ _____

2. $2SO_3$

3. H_2SO_3

4a. Two compounds

b. Two elements

c. Compound and element

5. Hg and O_2

6. NaCl and O_2

7. $CaCl_2$ and $2H_2O$

9. A third type of chemical reaction is called single replacement. Single replacement occurs when an element in a compound is replaced by a different element. Single replacement reactions have the following general form.

$$A + BC \longrightarrow AC + B$$

Note that the reactants consist of an element and a compound and the products consist of an element and a compound.

Example: $Zn + CuSO_4 \longrightarrow ZnSO_4 + Cu$

Zinc which can form a positive ion will replace the copper (II) ion which also forms a positive ion.

What will the products be for this reaction? Copper will form the copper (I) ion.

$$Cu + AgNO_3 \longrightarrow$$

---

10. In the last example, an element that could form a positive ion replaced a positive ion in a compound. An element that can form a negative ion can also replace a negative ion in a compound.

Example: $Cl_2 + 2NaBr \longrightarrow 2NaCl + Br_2$

What are the products of this reaction?

$$Br_2 + 2KI \longrightarrow$$

---

11. The fourth type of reaction is called double replacement. Double replacement occurs when positive ions in two different compounds change places. The general formula is:

$$AB + CD \longrightarrow AD + CB$$

Note that all of the reactants and products are compounds.

Example: $NaCl + AgNO_3 \longrightarrow AgCl + NaNO_3$

Silver and sodium have changed places. What will the products be in this reaction?

$$HCl + NaOH \longrightarrow$$

---

12. Identify these reactions as to type (combination, decomposition, etc.).

a. $Zn + 2HCl$ $\longrightarrow$ $ZnCl_2 + H_2$ _____

b. $CaO + H_2O$ $\longrightarrow$ $Ca(OH)_2$ _____

c. $2KClO_3$ $\longrightarrow$ $2KCl + 3O_2$ _____

d. $BaCl_2 + H_2SO_4$ $\longrightarrow$ $2HCl + BaSO_4$ _____

e. $CaCl_2 \cdot 2H_2O$ $\longrightarrow$ $CaCl_2 + 2H_2O$ _____

f. $Cl_2 + 2NaI$ $\longrightarrow$ $2NaCl + I_2$ _____

g. $4Fe + 3O_2$ $\longrightarrow$ $2Fe_2O_3$ _____

h. $Al_2(SO_4)_3 + 6NaOH$ $\longrightarrow$ $2Al(OH)_3 + 3Na_2SO_4$ _____

---

13. Answers to frame 12.

a. single replacement   b. combination   c. decomposition

d. double replacement   e. decomposition   f. single replacement

g. combination   h. double replacement

8a. Compound and element

b. Two elements

c. Two compounds

9. $CuNO_3$ and Ag

10. $2KBr$ and I_2

11. NaCl and H_2O

NAME _____

BALANCING CHEMICAL EQUATIONS
EVALUATION TEST 1
PART 1–PART 3

1. How many atoms of each type are represented in these formulas?

   $4Fe_2O_3$ a. Fe = _____   b. O = _____

   $3Ca(OH)_2$ c. Ca = _____   d. O = _____   e. H = _____

   $2Al_2(SO_4)_3$ f. Al = _____   g. S = _____   h. O = _____

2. Write the symbols of the seven diatomic elements.

   _____   _____   _____   _____   _____   _____   _____

3. Balance these chemical equations by the inspection method.

   a. $H_2 + Cl_2 \longrightarrow HCl$

   b. $HgO \longrightarrow Hg + O_2$

   c. $KClO_3 \longrightarrow KCl + O_2$

   d. $Na + CO_2 \longrightarrow C + Na_2O$

   e. $NaCl + Pb(NO_3)_2 \longrightarrow NaNO_3 + PbCl_2$

4. State whether the following elements have gained or lost electrons and how many.

	Gained or lost?	How many?
$Sn^0 \longrightarrow Sn^{4+}$	a. _____	b. _____
$Fe^{3+} \longrightarrow Fe^{2+}$	c. _____	d. _____
$N^{3-} \longrightarrow N^{5+}$	e. _____	f. _____

5. Oxidation involves the (a) _____ of electrons and reduction involves the

   (b) _____ of electrons.

BALANCING CHEMICAL EQUATIONS
EVALUATION TEST 2
PART 4

1. What elements in the equation below are free elements? _____

$$Na + H_2O \longrightarrow NaOH + H_2$$

2. What is the oxidation number of the monatomic ion of each of the following elements?

a. magnesium _____   b. sulfur _____   c. potassium _____

3. Assign oxidation numbers to the monatomic ions that make up the following compounds.

a. $MgCl_2$ _____

b. Na_2S _____

c. FeO _____

4. Assign oxidation numbers to hydrogen and oxygen in the equation below.

_____

$$H_2 + O_2 \longrightarrow H_2O$$

5. Assign oxidation numbers to the elements in these compounds.

a. HNO_3 _____

b. Na_2SO_4 _____

c. $KClO_4$ _____

6. Assign oxidation numbers to all the elements in these equations.

a. $H_2S + HNO_3 \longrightarrow S + NO + H_2O$

b. $KI + NaClO + H_2O \longrightarrow KOH + NaCl + I_2$

c. $KIO_4 + KI + HCl \longrightarrow KCl + I_2 + H_2O$

BALANCING CHEMICAL EQUATIONS
EVALUATION TEST 3
PART 5

1. Locate the elements that have lost or gained electrons in the equation below and connect them with lines.

$$P + HNO_3 + H_2O \longrightarrow NO + H_3PO_4$$

2. Determine the electron loss and gain in the following equation.

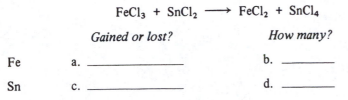

$$FeCl_3 + SnCl_2 \longrightarrow FeCl_2 + SnCl_4$$

	Gained or lost?		*How many?*
Fe	a. _____	b.	_____
Sn	c. _____	d.	_____

3. Balance these equations by the redox method.

a. $S + HNO_3 \longrightarrow H_2SO_4 + NO$

b. $H_2S + HNO_3 \longrightarrow S + NO + H_2O$

c. $HNO_3 + HI \longrightarrow NO + I_2 + H_2O$

d. $CaS + I_2 + HCl \longrightarrow CaCl_2 + HI + S$

e. $CuO + NH_3 \longrightarrow N_2 + H_2O + Cu$

BALANCING CHEMICAL EQUATIONS
EVALUATION TEST 4
PART 6 AND 7

Balance these redox equations written in ionic form.

1. $Zn + H^+ + NO_3^- \longrightarrow Zn^{2+} + N_2O + H_2O$

2. $MnO_4^- + S^{2-} + H_2O \longrightarrow MnO_2 + S + OH^-$

3. $Ca + H^+ \longrightarrow Ca^{2+} + H_2$

4. $Al + H^+ \longrightarrow Al^{3+} + H_2$

5. $NO_3^- + Br^- + H^+ \longrightarrow NO + Br_2 + H_2O$

Identify the type of reaction.

6. $2H_2O_2 \longrightarrow 2H_2O + 2O_2$ _____

7. $N_2 + 2O_2 \longrightarrow 2NO_2$ _____

8. $2Al + H_2SO_4 \longrightarrow Al_2SO_4 + H_2$ _____

9. $3Cu(OH)_2 + 2H_3PO_4 \longrightarrow Cu_3(PO_4)_2 + 6H_2O$ _____

10. $2BaO_2 \longrightarrow 2BaO + 2O_2$ _____

UNIT 11
Weight Relations in Chemistry

In this unit you will learn basic chemical measurement concepts and how to use them to solve problems. You will learn the concepts of atomic mass, atomic weight, and molecular weight, and the general mole concept. You will learn what isotopes are and what effect they have on atomic weight, how to use the concept of the mole as associated with atomic weight and as associated with molecular weight, and how to use the concept of the mole to solve problems. You will find the percentage composition and empirical and molecular formulas of compounds.

Part 1. Atomic Mass and Atomic Weight

1. The actual mass of an atom is very small. A hydrogen atom, for example, has a mass of only 1.67×10^{-24} grams. This is an inconveniently small number and not a very practical one to use. To simplify chemical measurements, chemists use a relative-mass system for the elements. The system is based on the mass of the carbon isotope known as carbon-12. This isotope has been arbitrarily assigned a value of 12 atomic-mass units. Thus, an atom with a mass twice as great as that of a carbon atom would have an atomic mass of 24.

   An atom with a mass of only one-half as much as a carbon atom would have an atomic mass of_____ .

2. The mass of an atom is based on the total number of protons and neutrons in the atom. The following chart shows the mass of the various subatomic particles.

Particle	Mass
proton	1
neutron	1
electron	0

   └─ (An electron actually has some mass, but it is so small that we can safely ignore it.)

   Using a carbon atom with an atomic mass of 12 as our example, the following chart shows how the number of protons and neutrons determines the mass. The carbon atom has six protons and six neutrons. Adding these up gives us the mass of the carbon atom.

1. 6

Particle	Number of Particles
proton	6
neutron	6
mass of carbon =	12

A proton has a mass of (a) _____ and a neutron has a mass of (b) _____ .

3. What is the mass of the following atom? _____

Particle	Number
protons	10
neutrons	12
mass =	?

4. How many neutrons does this atom have? _____

Particle	Number
protons	14
neutrons	?
mass =	30

5. How many protons does this atom have? _____

Particle	Number
protons	?
neutrons	22
mass = =	40

6. The mass is the number of (a)_____ plus the number of

(b) _____ .

2. a. one
   b. one

3. 22

4. 16

5. 18

6. a. protons
   b. neutrons

Part 2. Isotopes

1. Atoms of an element having the same number of protons–and therefore the same atomic number–but different numbers of neutrons–and hence different masses are called *isotopes*.

   Example: The three isotopes of hydrogen are as follows.

Particles	Isotope #1	Isotope #2	Isotope #3
protons	1	1	1
neutrons	0	1	2
mass	1	2	3

   It is very important to note that all of the isotopes of hydrogen have the *same* atomic number, or number of protons.

   Isotopes of an element have varying numbers of neutrons and thus have _____

   _____ masses.

2. Isotopes of an element have _____ atomic number(s).

3. Isotopes are caused by varying numbers of _____ in the nucleus of the atoms of an element.

4. We do not use the masses of the individual isotopes in chemistry. Instead, we average the mass of all the isotopes of an element, taking the abundance of each type of isotope into account. This average mass is then called the *atomic mass*. If you look at the periodic table of the elements on page 114, you will find the following information given for the element potassium.

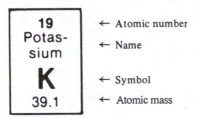

You are already familiar with the function of the atomic number. We are now interested in the atomic mass. Note that the atomic mass is not a whole number but a decimal fraction. The atomic mass of potassium shown on the table is the average mass of all the isotopes of potassium, taking the abundance of each type of isotope into account.

Find the atomic mass of the following elements, using the table.

a. Zn _____   b. Cl _____   c. Na _____

5. What elements have the following atomic masses?

a. 12.0 _____   b. 16.0 _____

c. 79.9 _____

6. Find the atomic mass of the elements with the following atomic numbers.

a. 7 _____   b. 16 _____   c. 13 _____

7. The atomic mass of the elements are decimal fractions because they represent the average mass of the _____.

Part 3. Atomic Mass and the Mole

1. The balances in a chemistry lab measure not atomic masses but grams. So a system has been developed for expressing atomic masses in grams, and we call this mass the molar mass.

Example:

Element	Atomic Mass	Molar Mass
Al	27.0	27.0 g
K	39.1	39.1 g

As you can see, the molar mass of an element is simply the number of grams of an element *numerically* equal to its atomic mass. (You have already learned that the atomic mass of an element is the average mass of the isotopes of that element, relative to the atomic mass of carbon-12.)

---

1. different

2. the same

3. neutrons

4. a. 65.4
   b. 35.5
   c. 23.0

5. a. carbon
   b. oxygen
   c. bromine

6. a. 14.0
   b. 32.1
   c. 27.0

7. isotopes

2. Another important concept in chemistry is a measure called the *mole*. In the discussion of atomic masses at the beginning of this chapter, you learned that the carbon isotope known as carbon-12 is used in determining the atomic-mass numbers of the elements. This same isotope is used to define the mole. A *mole* is an amount of a substance containing the same number of elementary units (atoms, molecules, electrons, or ions) as there are atoms in exactly 12 g of carbon-12. (The approximate number of atoms in 12 g of carbon-12 has been found to be 6.02×10^{23}, a number that is known as *Avogadro's Number.*

Avogadro's Number (6.02×10^{23}) of atoms = 1 mole of atoms = 1 molar mass

3. The mole is used to represent amounts of many particles, including atoms, ions, electrons, and molecules. Just as a *quart* of water represents 32 oz of water, a *mole* of atoms represents 6.02×10^{23} atoms. The example below shows the relationship between the molar mass and the mole.

Example:

Element	Molar Mass	Number of Moles
Al	27.0 g	1
K	39.1 g	1

The weight of a mole is dependent on the weight of the substance being measured, because the mole can be a mole of atoms, a mole of molecules, a mole of electrons, and so on.

Example:

Element	Atomic Mass	Molar Mass
Al	27.0	27.0 g
K	39.1	39.1 g

Find the weight of 1 mole of these elements.

Element	Atomic Mass	Molar Mass
Na	a. _____	b. _____
Br	c. _____	d. _____

4. What is the Molar mass of these elements?

   a.   sulfur _____

   b.   magnesium _____

5. The weight of 1 mole is equivalent to the atomic mass expressed in _____ .

6. It is often necessary to know the number of moles of atoms represented by a given mass of an element. We use a mathematical equation to arrive at the number of moles. The steps involved in using the equation are listed below. For convenience we round off atomic masses to one decimal place.

Step 1. Write the formula of the math equation:

$$\text{number of moles} = \frac{\text{given weight (of the element) in grams}}{\text{weight of 1 mole (molar mass)}}$$

Step 2. List the data needed to solve the equation.
Step 3. Substitute the data into the formula.
Step 4. Do the calculation indicated.

3. a. 23.0
   b. 23.0 g
   c. 79.9
   d. 79.9 g

4. a. 32.1 g
   b. 24.3 g

5. grams

Example: How many moles of potassium does 78.2 g of potassium represent?

Step 1. Write the formula.

$$\text{number of moles} = \frac{\text{given weight (of the element) in grams}}{\text{weight of 1 mole (molar mass)}}$$

Step 2. List the data.

given weight in grams = 78.2 g
weight of 1 mole = 39.1 g

Step 3. Substitute the data into the formula.

$$\text{number of moles} = \frac{78.2 \text{ g}}{39.1 \text{ g}}$$

Step 4. Do the calculation.

number of moles = 2.00

Determine the number of moles represented by 12.2 g of magnesium.

Step 1.

Step 2.

Step 3.

Step 4.

---

7. Answers to frame 6.

Step 1. Write the formula.

$$\text{number of moles} = \frac{\text{given weight in grams}}{\text{weight of 1 mole}}$$

Step 2. List the data.

given weight in grams = 12.2 g
weight of 1 mole = 24.3 g

Step 3. Substitute the data into the formula.

$$\text{number of moles} = \frac{12.2 \text{ g}}{24.3 \text{ g}}$$

Step 4. Do the calculation.

number of moles = 0.502

---

8. List the four steps involved in solving the mathematical equation for determining the number of moles of atoms represented by a given weight of an element.

Step 1. _____

Step 2. _____

Step 3. _____

Step 4. _____

9. Answers to frame 8.

Step 1. Write the formula.   Step 2. List the data.
Step 3. Substitute the data into the formula.   Step 4. Do the calculation.

---

10. How many moles of calcium atoms are represented by 120.3 g of calcium? (Use all four steps.)

10. 3.00 moles

---

11. Conversion factors can be used to solve this type of problem.

Example: How many moles of Fe does 25.0g of Fe represent?

Step 1. Set up the conversion factor changing grams to moles.

$$\text{grams Fe} \times \frac{1 \text{ mole Fe}}{\text{Molar mass Fe}} \quad \text{(wt. per mole)}$$

Step 2. List the data.
grams Fe = 25.0g
molar mass Fe = 55.9g

Step 3. Substitute into the formula and solve.
$$25.0\text{g Fe} \times \frac{1 \text{ mole Fe}}{55.8\text{g Fe}} = 0.448 \text{ mole Fe}$$

---

12. How many moles of Cu does 30.0g of Cu represent?

---

13. Answer to frame 12.

1. $\text{grams Cu} \times \dfrac{1 \text{ mole Cu}}{\text{molar mass Cu}}$

2. grams Cu = 30.0g
molar mass Cu = 63.6g

3. $30.0\text{g} \times \dfrac{1 \text{ mole Cu}}{63.5\text{g Cu}} = 0.472 \text{ mole Cu}$

---

14. Find the number of moles in the following.
   a. 22.0g of C   b. 40.5g of Al

---

15. Answer to frame 14.
   a. $22.0\text{g C} \times \dfrac{1 \text{ mole C}}{12.0\text{g C}} = 1.83 \text{ mole C}$

   b. $40.5\text{g Al} \times \dfrac{1 \text{ mole Al}}{27.0\text{g Al}} = 1.50 \text{ mole Al}$

Part 4. Molar Mass of a Compound

1. Just as we can find the atomic mass of an element, we can find the molecular mass of a compound. The molecular mass of a compound is found by adding up all of the atomic masses in the compound.

    Example: What is the molecular mass of H_2O?

Type of Atom	Atomic Mass		Number of Atoms in Formula		Molecular Mass
H	1.0	X	2	=	2.0
O	16.0	X	1	=	16.0
					18.0

    Find the molecular mass of $Al(OH)_3$.

Type of Atom	Atomic Mass		Number of Atoms		Molecular Mass
Al	a. _____	X	b. _____	=	c. _____
O	d. _____	X	e. _____	=	f. _____
H	g. _____	X	h. _____	=	i. _____
					j. _____

2. Answers to frame 1.

a. 27.0	b. 1	c. 27.0	d. 16.0	e. 3
f. 48.0	g. 1.0	h. 3	i. 3.0	j. 78.0

3. What is the molecular mass of $Ca_3(PO_4)_2$?

Part 5. Molecular Mass and the Mole

1. We can express molecular mass in grams just as we express atomic mass in grams. And just as 1 molar mass is equivalent to 1 mole of atoms of an element, 1 molar mass is equivalent to 1 mole of molecules of a compound.

    3. 310.3

    Example: H_2O

Type of Atom	Atomic Mass		Number of Atoms		Molecular Mass		Molar Mass		Weight of 1 Mole of H_2O
H	1.0	X	2	=	2.0	=	2.0 g		
O	16.0	X	1	=	16.0	=	16.0 g		
					18.0	=	18.0 g	=	18.0 g

Here is another example, omitting the steps used to arrive at the molecular mass.

Example:

Compound	*Molecular Mass*		*Molar Mass*		*Weight of 1 Mole of CO_2*
CO_2	44.0	=	44.0 g	=	44.0 g

What is the weight of 1 mole of these compounds?

Compound	*Molecular Mass*	*Weight of 1 Mole*
NO_2	a. _____	b. _____
NH_3	c. _____	d. _____

2. What is the weight of 1 mole of these compounds?

   a. CH_4 _____   b.   NaOH _____

3. 1 mole of atoms is equivalent to the _____ expressed in

   grams.

4. 1 mole of molecules is equivalent to the _____

   expressed in grams.

5. It is often necessary to know the number of moles represented by the given weight of a compound. A mathematical equation is used to determine this number. The steps involved in using the equation are listed below.

   Example: How many moles of CO_2 are represented by 88.0 g of CO_2?

   Step 1.   Write the formula of the equation.

   $$\text{number of moles} = \frac{\text{given weight in grams}}{\text{weight of 1 mole (of the compound)}}$$

   Step 2.   List the data needed to solve the formula.

   given weight in grams = 88.0 g
   weight of 1 mole of C = 12.0 g × 1 = 12.0 g
   weight of 1 mole of O = 16.0 g × 2 = $\underline{32.0 \text{ g}}$
                                             44.0 g

   Step 3.   Substitute the data into the formula.

   $$\text{number of moles} = \frac{88.0 \text{ g}}{44.0 \text{ g}}$$

   Step 4.   Do the calculation.

   number of moles = 2.00

How many moles are represented by 9.0 g of water?

Step 1.   Write the formula.

---

1. a. 46.0
   b. 46.0 g
   c. 17.0
   d. 17.0 g

2. a. 16.0 g
   b. 40.0 g

3. atomic mass

4. molecular mass

Step 2. List the data.

Step 3. Substitute the data into the formula.

Step 4. Do the calculation.

---

6. Answers to frame 5.

Step 1. number of moles $= \dfrac{\text{given weight in grams}}{\text{weight of 1 mole}}$

Step 2. given weight in grams $= 9.0$ g
  weight of 1 mole of H $= 1.0 \times 2 = 2.0$ g
  weight of 1 mole of O $= 16.0 \times 1 = \underline{16.0\ \text{g}}$
  18.0 g

Step 3. number of moles $= \dfrac{9.0\ \text{g}}{18.0\ \text{g}}$

Step 4. number of moles $= 0.50$

---

7. How many moles does 138.0 g of NO_2 represent?

---

8. Answers to frame 7.

Step 1. number of moles $= \dfrac{\text{weight in grams}}{\text{weight of 1 mole}}$

Step 2. weight in grams $= 138.0$
  weight of 1 mole of N $= 14.0 \times 1 = 14.0$ g
  weight of 1 mole of O $= 16.0 \times 2 = \underline{32.0\ \text{g}}$
  46.0 g

Step 3. number of moles $= \dfrac{138.0\ \text{g}}{46.0\ \text{g}}$

Step 4. number of moles $= 3.00$

---

9. The atomic mass and the molecular mass of a diatomic element are two different quantities. In one molecular mass of a diatomic element, there are two atomic masses of that element.

Example: What is the atomic and molecular mass of hydrogen?

Element	Atomic Mass	Molecular Mass
H_2	1.0	$1.0 \times 2 = 2.0$

What are the atomic and molecular masses of the listed elements?

a. O_2 b. N_2 c. Cl_2 d. I_2

10. Answers to frame 9.

a. Atomic mass of O = 16.0
Molecular mass of O_2 = 32.0
c. Atomic mass of Cl = 35.5
Molecular mass of Cl_2 = 71.0

b. Atomic mass of N = 14.0
Molecular mass of N_2 = 28.0
d. Atomic mass of I = 126.9
Molecular mass of I_2 = 253.8

11. Conversion factors can be used to solve for the number of moles of a compound.

Example: How many moles of NaOH are there in 100g of NaOH?

Step 1. Set up the conversion factor changing grams to moles.

$$\text{grams NaOH} \times \frac{1 \text{ mole NaOH}}{\text{Molar mass NaOH}}$$

Step 2. List the data.
grams NaOH = 100g
molar mass NaOH = Na 23.0g
                   O 16.0g
                   H   1.0g
                      40.0g

Step 3. Substitute into the formula and solve.

$$100\text{g NaOH} \times \frac{1 \text{ mole NaOH}}{40.0\text{g NaOH}} = 2.50 \text{ moles NaOH}$$

12. Use conversion factors to find how many moles of H_2O are in 6.00g of H_2O.

13. Answer to frame 12.

1. Grams $H_2O \times \dfrac{1 \text{ mole } H_2O}{\text{molar mass } H_2O}$

2. Grams H_2O = 6.00g
molar mass H_2O = 18.0g

3. 6.00 Grams $H_2O \times \dfrac{1 \text{ mole } H_2O}{18.0\text{g } H_2O} = 0.333$ Mole H_2O

14. Use conversion factors to find the number of moles of NH_3 in 42.5 g of NH_3.

15. Answer to frame 14.

$$42.5 \text{g NH}_3 \times \frac{1 \text{ mole NH}_3}{17.0 \text{g NH}_3} = 2.50 \text{ mole NH}_3$$

Part 6. The General Mole Concept

1. The term *mole* is used in chemistry to describe the same number of units of atoms, molecules, formula units, ions, or electrons. The number of units in a mole is Avogadro's Number (6.02×10^{23}).

1 mole of atoms	= 1 molar mass	= 6.02×10^{23} atoms
1 mole of molecules	= 1 molar mass	= 6.02×10^{23} molecules
1 mole of formula units	= 1 mass formula	= 6.02×10^{23} formula units
1 mole of ions	= 1 mass ionic	= 6.02×10^{23} ions

A mole of any substance is 6.02×10^{23} units of that substance.

Another formula can be added to the list of ways to calculate the number of moles of a substance.

Formulas for Calculating the Number of Moles of a Substance

1. Number of moles of atoms $= \dfrac{\text{wt in grams of the atom}}{\text{wt of 1 mole (of the atom)}}$

2. Number of moles of molecules $= \dfrac{\text{wt in grams of the molecule}}{\text{wt of 1 mole (of the molecule)}}$

3. Number of moles of any substance $= \dfrac{\text{Number of units of substance}}{6.02 \times 10^{23} \text{ units/mole}}$

Here are examples using formula three.

A. If you have 18.06×10^{23} molecules of a substance, how many moles would you have?

1. Number of moles $= \dfrac{\text{Number of molecules}}{6.02 \times 10^{23} \text{ molecules/mole}}$

2. Number of units $= 18.06 \times 10^{23}$ molecules
   Number of units/mole $= 6.02 \times 10^{23}$ molecules/mole

3. Number of moles $= \dfrac{18.06 \times 10^{23} \text{ molecules}}{6.02 \times 10^{23} \text{ molecules/mole}}$

4. Number of moles $= 3.00$ moles

B. If you have 2.50 moles of ions, how many ions would you have? Use a conversion factor.

$$\text{Number of ions} = \text{Number of moles} \times \frac{6.02 \times 10^{23} \text{ ions}}{1 \text{ mole}}$$

$$\text{Number of ions} = 2.50 \text{ moles} \times \frac{6.02 \times 10^{23} \text{ ions}}{1 \text{ mole}}$$

$$\text{Number of ions} = 15.1 \times 10^{23} \text{ or } 1.51 \times 10^{24}$$

Try these problems.

Calculate how many moles are represented by the given number of units.

a. 3.01×10^{23} atoms

b. 9.03×10^{23} electrons

Calculate how many molecules are represented.

c. 4.20 moles of a substance

d. 0.455 moles of a substance

---

2. Answers to frame 1.

a. no. of moles $= \dfrac{3.01 \times 10^{23} \text{ atoms}}{6.02 \times 10^{23} \text{ atoms/mole}} = 0.500$ moles

b. no. of moles $= \dfrac{9.03 \times 10^{23} \text{ electrons}}{6.02 \times 10^{23} \text{ electrons/mole}} = 1.50$ moles

c. 4.20 moles $\times \dfrac{6.02 \times 10^{23} \text{ molecules}}{1 \text{ mole}} = 25.3 \times 10^{23}$ molecules,

or 2.53×10^{24} molecules

d. 0.455 moles $\times \dfrac{6.02 \times 10^{23} \text{ molecules}}{1 \text{ mole}} = 2.74 \times 10^{23}$ molecules

---

3. Solve these problems.

How many neutrons do these atoms have?

a. protons   42
    neutrons   —
    mass number 86

b. protons   16
    neutrons   —
    mass number 30

Determine the number of moles represented by the given weight in grams of the following elements.

c. 63.0 g of iron _____

d. 4.50 g of sulfur _____

e. 100 g of calcium _____

f. 11.2 g of magnesium _____

Calculate the molecular mass of the following compounds.

g. $Al_2(CO_3)_3$ _____

h. $(NH_4)_2SO_4$ _____

i. $KClO_3$ _____

J. $Al_2(SO_4)_3$ _____

Determine the number of moles represented by the given weight in grams of the following compounds.

k. 13.5 g of H_2O _____

l. 66 g of CO_2 _____

m. 39.1 g of NH_3 _____

n. 25.0 g of NaOH _____

What are the atomic and molecular masses of the listed elements?

o. Br_2 _____   p. F_2 _____

q. How many moles are represented by 21.07×10^{23} ions?

r. How many molecules are there in 0.617 moles of a substance?

---

4. Answers to frame 3.

a. 44   b. 14   c. 1.13 moles   d. 0.140 moles
e. 2.49 moles   f. 0.461 moles   g. 234.0   h. 132.1
i. 122.6   j. 342.3   k. 0.750 moles   l. 1.5 moles
m. 2.30 moles   n. 0.625 moles   o. 79.9 and 159.8   p. 19.0 and 38.0
q. 3.5 moles   r. 3.71×10^{23} molecules

Part 7. Percentage Composition of Compounds, Empirical Formula, and Molecular Formula

1. The formula mass or the molecular mass of a compound represents the total weight of that compound. We can calculate what fraction of the total weight each element in that compound represents. If this fractional weight of each element is expressed as a percentage, we have the percentage composition of a compound. The term mass and weight are interchangeable.

Example: Calculate the percentage composition of CO_2.

Step 1.   Calculate the molecular mass of CO_2.

$$C: 12.0 \times 1 = 12.0$$
$$O: 16.0 \times 2 = 32.0$$
$$44.0$$

C: $12.0 \times 1 = 12.0$ → fractional weights
O: $16.0 \times 2 = 32.0$ →
44.0 ← total weight

Step 2. Divide each fractional weight by the total weight and multiply by 100 to get the percentage composition of each element.

C: $\dfrac{12.0}{44.0} \times 100 = 27.3\%$ carbon

O: $\dfrac{32.0}{44.0} \times 100 = 73.7\%$ oxygen

The percentage should add up to 100. Check your answers by doing this step. You may not always get 100% exactly because of rounding off.

Calculate the percentage composition of $NaNO_3$.

Step 1.

Step 2.

2. Answers to frame 1.

Step 1.

Na: $23.0 \times 1 = 23.0$
N: $14.0 \times 1 = 14.0$
O: $16.0 \times 3 = 48.0$
$\overline{85.0}$

Step 2.

Na: $\dfrac{23.0}{85.0} \times 100 = 27.1\%$

N: $\dfrac{14.0}{85.0} \times 100 = 16.5\%$

O: $\dfrac{48.0}{85.0} \times 100 = 56.5\%$

---

3. Calculate the percentage composition of these compounds.

   a. FeO   b. NaCl   c. $MgSO_4$ d. $Ca_3(PO_4)_2$

---

4. Answers to frame 3.

   a. Fe: $\dfrac{55.8}{71.8} \times 100 = 77.7\%$ b. Na: $\dfrac{23.0}{58.5} \times 100 = 39.3\%$

       O: $\dfrac{16}{71.8} \times 100 = 22.3\%$ Cl: $\dfrac{35.5}{58.5} \times 100 = 60.7\%$

   c. Mg: $\dfrac{24.3}{120.4} \times 100 = 20.2\%$ d. Ca: $\dfrac{120.3}{310.3} \times 100 = 38.77\%$

       S: $\dfrac{32.1}{120.4} \times 100 = 26.7\%$ P: $\dfrac{62.00}{310.3} \times 100 = 19.98\%$

       O: $\dfrac{64.0}{120.4} \times 100 = 53.2\%$ O: $\dfrac{128.0}{310.3} \times 100 = 41.25\%$

---

5. Given the formula of a compound, you can calculate the weight percentage of each element. The reverse can also be done: Given the weight percentage of each element in a compound, you can calculate the empirical formula of a compound. The empirical formula gives the lowest ratio of the atoms that are present in the compound.

Example: Calculate the empirical formula of a compound containing 11.19% hydrogen and 88.89% oxygen.

Step 1.   The term *percentage* means parts per 100. This means that, if we have 11.19% hydrogen, 11.19 g out of every 100 g of the compound is hydrogen. If we have 88.89% of oxygen, 88.89 g out of every 100 g of the compound is oxygen.

H: $11.19\% = 11.19/100$ or 11.19 g
O: $88.89\% = 88.89/100$ or 88.89 g

The purpose of Step 1 is to change the percentage weight to grams.

Step 2.   In this step the grams are changed to moles. Use the conversion factor:

$$\text{Number of moles} = \frac{\text{given weight in grams}}{\text{weight of one mole of an atom}}$$

H: number of moles $= \dfrac{11.19 \text{ g}}{1.01 \text{ g/ mole}} = 11.1$ moles

O: number of moles $= \dfrac{88.89 \text{ g}}{16.0 \text{ g/ mole}} = 5.55$ moles

The empirical formula $= H_{11.1}O_{5.55}$

We could stop at this point, but we need whole numbers for subscripts in a formula. This is accomplished in Step 3.

Step 3.   Divide each quantity in Step 2 by the smaller of the quantities (5.55 moles of oxygen).

H: $\dfrac{11.1 \text{ moles}}{5.55 \text{ moles}} = 2$

O: $\dfrac{5.55 \text{ moles}}{5.55 \text{ moles}} = 1$

The empirical formula is H_2O.

Here is another example. What is the empirical formula of a compound that contains 63.53% iron and 36.47% sulfur?

Step 1.   Fe: 63.53% = 63.53/100 = 63.53 g
           S: 36.47% = 36.47/100 = 36.47 g

Step 2.   Number of moles $= \dfrac{\text{given wt in grams}}{\text{weight of one mole}}$

Fe: Number of moles $= \dfrac{63.53 \text{ g}}{55.8 \text{ g/mole}} = 1.14$ moles

S: Number of moles $= \dfrac{36.47 \text{ g}}{32.1 \text{ g/mole}} = 1.14$ moles

Step 3.   Fe: $\dfrac{1.14 \text{ mole}}{1.14 \text{ mole}} = 1$

S: $\dfrac{1.14 \text{ mole}}{1.14 \text{ mole}} = 1$

The empirical formula is FeS.

Try this problem.

Calculate the empirical formula of a compound that contains 31.9% K, 28.9% Cl, and 39.2% O.

Step 1.

Step 2.

Step 3.

---

6. Answers to frame 5.

Step 1.
           K: 31.9/100 = 31.9 g
           Cl: 28.9/100 = 28.9 g
           O: 39.2/100 = 39.2 g

Step 3.

K: $\dfrac{0.816 \text{ mole}}{0.815 \text{ mole}} = 1.00$

Cl: $\dfrac{0.815 \text{ mole}}{0.815 \text{ mole}} = 1.00$

O: $\dfrac{2.45 \text{ mole}}{0.815 \text{ mole}} = 3.01$

Step 2.

K: $\dfrac{31.9 \text{ g}}{39.1 \text{ g/mole}} = 0.816$ mole

Cl: $\dfrac{28.9 \text{ g}}{35.5 \text{ g/mole}} = 0.815$ mole

O: $\dfrac{39.2 \text{ g}}{16.0 \text{ g/mole}} = 2.45$ moles

The formula is $KClO_3$.

7. Determine the empirical formula for the compounds listed, given their percentage composition.

   a. 77.7% Fe and 22.3% O

   b. 27.3% C and 72.7% O

   c. 40.3% K, 26.7% Cr, and 33.0% O

   d. 56.58% K, 8.68% C, and 34.73% O

8. Answers to frame 7.

   a. Fe: $\dfrac{77.7 \text{ g}}{55.8 \text{ g/mole}}$ = 1.39 mole, $\dfrac{1.39}{1.39}$ = 1.00

       O: $\dfrac{22.3 \text{ g}}{16.0 \text{ g/mole}}$ = 1.39 mole, $\dfrac{1.39}{1.39}$ = 1.00

       The formula is FeO.

   b. C: $\dfrac{27.3 \text{ g}}{12.0 \text{ g/mole}}$ = 2.28 mole, $\dfrac{2.28}{2.28}$ = 1.00

       O: $\dfrac{72.7 \text{ g}}{16.0 \text{ g/mole}}$ = 4.54 mole, $\dfrac{4.54}{2.28}$ = 1.99

       The formula is CO_2.

   c. K: $\dfrac{40.3 \text{ g}}{39.1 \text{ g/mole}}$ = 1.03 mole, $\dfrac{1.03}{0.513}$ = 2.01

       Cr: $\dfrac{26.7 \text{ g}}{52.0 \text{ g/mole}}$ = 0.513 mole, $\dfrac{0.513}{0.513}$ = 1.00

       O: $\dfrac{33.0 \text{ g}}{16.0 \text{ g/mole}}$ = 2.06 mole, $\dfrac{2.06}{0.513}$ = 4.02

       The formula is K_2CrO_4.

   d. K: $\dfrac{56.58 \text{ g}}{39.1 \text{ g/mole}}$ = 1.45 mole, $\dfrac{1.45}{0.723}$ = 2.01

       C: $\dfrac{8.68 \text{ g}}{12.0 \text{ g/mole}}$ = 0.723 mole, $\dfrac{0.723}{0.723}$ = 1.00

       O: $\dfrac{34.73 \text{ g}}{16.0 \text{ g/mole}}$ = 2.17 mole, $\dfrac{2.17}{0.723}$ = 3.00

       The formula is K_2CO_3.

9. At times, an extra step is needed to obtain whole numbers for subscripts.

   Example: Calculate the empirical formula for a compound containing 69.9% Fe and 30.1% O.

   Step 1.   Fe: 69.9% = 69.9/100 = 69.9 g
   O: 30.1% = 30.1/100 = 30.1 g

   Step 2.   Fe: number of moles = $\dfrac{69.9 \text{ g}}{55.8 \text{ g/mole}}$ = 1.25 moles

   O: number of moles = $\dfrac{30.1 \text{ g}}{16.0 \text{ g/mole}}$ = 1.88 moles

   Step 3.   Fe: $\dfrac{1.25 \text{ moles}}{1.25 \text{ moles}}$ = 1.00

   O: $\dfrac{1.88 \text{ moles}}{1.25 \text{ moles}}$ = 1.50

   Step 4.   Multiply the numbers in Step 3 by a number that will give you whole numbers.

   Fe: 1.00 X 2 = 2.00
   O: 1.50 X 2 = 3.00

   The empirical formula is Fe_2O_3.

Calculate the empirical formula of these compounds given their percentage composition.

a. 26.57% K, 35.36% Cr, and 38.07% O

b. 18.4% Al, 32.6% S, and 49.0% O

10. Answers to frame 9.

   a.   K: $\dfrac{26.57 \text{ g}}{39.1 \text{ g/mole}}$ = 0.680 mole, $\dfrac{0.680}{0.680}$ = 1.00, 1.00 X 2 = 2.00

   Cr: $\dfrac{35.36 \text{ g}}{52.0 \text{ g/mole}}$ = 0.680 mole, $\dfrac{0.680}{0.680}$ = 1.00, 1.00 X 2 = 2.00

   O: $\dfrac{38.07 \text{ g}}{16.0 \text{ g/mole}}$ = 2.38 mole, $\dfrac{2.38}{0.680}$ = 3.50, 3.50 X 2 = 7.00

   The formula is $K_2Cr_2O_7$.

   b.   Al: $\dfrac{18.4 \text{ g}}{27.0 \text{ g/mole}}$ = 0.681 mole, $\dfrac{0.681}{0.681}$ = 1.00, 1.00 X 2 = 2.00

   S: $\dfrac{32.6 \text{ g}}{32.1 \text{ g/mole}}$ = 1.02 mole, $\dfrac{1.02}{0.681}$ = 1.50, 1.50 X 2 = 3.00

   O: $\dfrac{49.0 \text{ g}}{16.0 \text{ g/mole}}$ = 3.06 mole, $\dfrac{3.06}{0.681}$ = 4.49, 4.49 X 2 = 8.98

   The formula is $Al_2(SO_3)_3$.

11. Several compounds can have the same empirical formula.

Here are examples of compounds with different molecular formulas but the same empirical formula.

Molecular Formula	Empirical Formula
C_2H_4	CH_2
C_3H_6	CH_2
C_4H_8	CH_2

The molecular formula, unlike the empirical formula, does not always have the lowest ratio for the subscripts. If the molecular and the empirical formulas are different, the subscripts of the molecular formula will be a whole-number multiple of the subscripts of the empirical formula.

Empirical Formula	Molecular Formula
$(CH_2)_n$ If $n = 2$	$(CH_2)_2 = C_2H_4$

The molecular mass of the compound must be known in order to calculate the molecular formula from the empirical formula.

Example: Calculate the molecular formula for a compound containing 14.3% H and 85.7% C. The molecular mass of the compound is 28.0 g/mole.

Part A.   First calculate the empirical mass.

Step 1. H $= 14.3\% = 14.3/100 = 14.3$ g
C $= 85.7\% = 85.7/100 = 85.7$ g

Step 2. H: number of moles $= \dfrac{14.3 \text{ g}}{1.01 \text{ g/mole}} = 14.2$ moles

C: number of moles $= \dfrac{85.7 \text{ g}}{12.0 \text{ g/mole}} = 7.14$ moles

Step 3. H $= \dfrac{14.2 \text{ moles}}{7.14 \text{ moles}} = 1.99$ (2.00, rounded)

C $= \dfrac{7.14 \text{ moles}}{7.14 \text{ moles}} = 1.00$

The empirical formula is CH_2.

Part B.   Calculate the molecular formula from the empirical formula and the molecular mass of the compound.

Step 1. Calculate the molecular mass of the empirical formula (CH_2).

C: $12.0 \times 1 = 12.0$ g
H: $1.0 \times 2 = \underline{2.0 \text{ g}}$
14.0 g/mole

Step 2. Divide the molecular mass of the molecular formula by the molecular mass of the empirical formula to obtain the multiple n.

$(CH_2)_n$ $n = \dfrac{28.0 \text{ g/mole}}{14.0 \text{ g/mole}} = 2$

The true or molecular formula is:

$(CH_2)_2$ or C_2H_4

Calculate the true or molecular formula of the following compounds.

a. 93.30% C, 7.76% H, and a molecular mass of 78.0 g/mole.

b. 5.88% H, 94.11% O, and a molecular mass of 34.0 g/mole.

---

12. Answers to frame 11.

a. C: $\dfrac{93.30 \text{ g}}{12.0 \text{ g/mole}}$ = 7.78 mole, $\dfrac{7.78}{7.76}$ = 1.00

   H: $\dfrac{7.76 \text{ g}}{1.00 \text{ g/mole}}$ = 7.76 mole, $\dfrac{7.76}{7.76}$ = 1.00

   The empirical formula is CH.
   The molecular mass of the molecular formula is 78.0 g.
   The molecular mass of the empirical formula is 13.0 g.

   $n = \dfrac{78.0 \text{ g}}{13.0 \text{ g}}$ = 6   The molecular formula is $(CH)_6$ = C_6H_6.

b. H: $\dfrac{5.88 \text{ g}}{1.0 \text{ g/mole}}$ = 5.88 mole, $\dfrac{5.88}{5.88}$ = 1.00

   O: $\dfrac{94.11 \text{ g}}{16.0 \text{ g/mole}}$ = 5.88 mole, $\dfrac{5.88}{5.88}$ = 1.00

   The empirical formula is HO.
   The molecular mass of the molecular formula is 34.0 g.
   The molecular mass of the empirical formula is 17.0 g.

   $n = \dfrac{34.0 \text{ g}}{17.0 \text{ g}}$ = 2   The molecular formula is $(HO)_2$ = H_2O_2.

---

13. Solve these problems.

a. Calculate the percentage composition of Al_2SO_4.

b. Calculate the percentage composition of $KClO_3$.

Determine the empirical formula for the compounds listed, given their percentage composition.

c. 40.1% Ca, 12.0% C, and 48.0% O

d. 3.1% H, 31.6% P, and 65.3% O

e. 43.7% P and 56.3% O

Calculate the molecular formula of the following compounds.

f. 14.4% H, 85.6% C, and a molecular mass of 42.0 g/mole.

g. 7.7% H, 92.3% C, and a molecular mass of 78.0 g/mole.

---

14. Answers to frame 13.

a. Al: 36.0%, S: 21.4%, O: 42.6% b. K: 31.9%, Cl: 28.9%, O: 39.2%
c. $CaCO_3$ d. H_3PO_4 e. P_2O_5 f. $(CH_2)_3 = C_3H_6$ g. $(CH)_6 = C_6H_6$

Part 8. The Percentage Composition of a Hydrated Compound

1. Some compounds have loosely bonded water molecules attached to them. The water present is called water of hydration. The water is easily driven off by heating. The percentage of water in the hydrated compound can be found in the following manner.

   Example: The hydrated compound $CuCl_2 \cdot 2H_2O$ is heated to yield $CuCl_2$ residue. The water is driven off as a gas. Find the percentage weight of the compound that is water if 0.950 g of the hydrated compound yields 0.750 g of residue.

   $$CuCl_2 \cdot 2H_2O \quad \rightarrow \quad CuCl_2 \quad + \quad 2H_2O$$
   $$0.950 \text{ g} \qquad\qquad 0.750 \text{ g} \qquad\qquad ?$$

   Step 1. Subtract the weight of the residue from the weight of the hydrated compound to obtain the weight of water driven off.

   0.950 g − 0.750 g = 0.200 g of water

   Step 2. Divide the weight of the water by the weight of the hydrated compound and multiply by 100 to find the percentage composition of water in the hydrated compound.

   $$\% \text{ of water} = \frac{0.200 \text{ g}}{0.950} \times 100$$

   % of water = 21.1%

Try this problem.

Hydrated $CoCl_2 \cdot 2H_2O$ is heated to produce $CoCl_2$ residue and water. If 0.825 g of the hydrated compound produces 0.646 g of residue, what is the percentage by weight of water in the hydrated compound?

---

2. Answer to frame 1.

    1. Weight of water

       0.825 g − 0.646 g = 0.179 g

    2. % of water

$$\% \text{ of water } = \frac{0.179 \text{ g}}{0.825 \text{ g}} \times 100 = 21.7\%$$

---

3. 0.750 g of $CuSO_4 \cdot 5H_2O$ are heated to produce 0.479 g of $CuSO_4$ residue. What is the percentage weight of water in the hydrated compound?

---

4. Answer to frame 3.

    1. 0.750 g − 0.479 g = 0.271 g

    2. $\% \text{ of water } = \dfrac{0.271 \text{ g}}{0.750 \text{ g}} \times 100 = 36.1\%$

WEIGHT RELATIONS IN CHEMISTRY
EVALUATION TEST 1
PART 1–PART 4

1. How many neutrons does this atom have?

Particle	Number
protons	20
neutrons	_____
mass number	42

2. Isotopes of an element have the same atomic (a) _____ but dif-
    ferent (b) _____

3. What is the atomic mass, to one decimal place, of these elements?

    a. Mg _____   b. Cu _____   c. S _____

4. List the four steps used in determining the number of moles of atoms represented
    by a given weight of an element.

    a. _____

    b. _____

    c. _____

    d. _____

5. Determine the number of moles represented by 80.25 g of sulfur. *Show all steps.*

6. Calculate the molar mass of $(NH_4)_3PO_4$.

WEIGHT RELATIONS IN CHEMISTRY
EVALUATION TEST 2
PART 5–PART 8

1. 110 g of MgO equal how many moles of MgO?

2. What are the atomic and molecular masses of N_2?

3. How many moles are represented by 30.1×10^{23} atoms?

4. How many molecules are there in 1.17 moles of a substance?

5. Calculate the percentage composition of $NaClO_3$.

6. Determine the empirical formula of the listed compound from its percentage composition.

    22.1% Al, 25.4% P, and 52.5% O

7. Determine the molecular formula of the listed compound from its percentage composition and molecular mass.

    85.7% C, 14.3% H, and a molecular mass of 84.0 g/mole

8. 0.950 g of $CoCl_2 \cdot 6H_2O$ are heated to produce 0.519 g of $CoCl_2$ residue. What is the percentage weight of water in the hydrated compound?

UNIT 12
Chemical Equations and Stoichiometry

In this unit you will combine your knowledge of writing equations and weight relations in chemistry to solve stoichiometric problems. You will make mole/ mole, mole/weight, weight/weight, and limiting-reagent calculations.

Part 1. Mole/Mole Calculations: The Mole-Ratio Method

1. What you have learned about balancing equations and what you have learned about finding the weight of a mole can be used to solve a problem in chemistry called the *mole/mole calculation*. This is a useful calculation that allows us to predict in advance what weight of product will be formed from a given weight of reactants.

   Example: $H_2 + Cl_2 \longrightarrow 2HCl$
   1 mole   1 mole   2 moles

   Note that the coefficients in front of the formulas tell us the ratio in which the reactants react and the quantity in which the products form.

   In the blanks below each formula, tell how many moles are represented.

   $2H_2 \qquad + \qquad O_2 \qquad \longrightarrow \qquad 2H_2O$

   a. _____   b. _____   c. _____

2. The ratios of reactants to products that are expressed in a balanced equation are constant. Once we have one constant relationship, we can use it to calculate any other relationship.

   Example: $H_2 + Cl_2 \longrightarrow 2HCl$
   1 mole   2 moles

   We can see that for every 1 mole of H_2 reacting with Cl we get 2 moles of HCl product. If we had twice as much H_2 reacting with Cl, we would get twice as much HCl. Two moles of H_2 will produce 4 moles of HCl in a balanced equation.

1. a.   2
   b.   1
   c.   2

3. The mathematical calculation of the quantitative relationship between reactants and products in a chemical reaction—which is called a *stoichiometric* problem—requires the use of a process called the *mole-ratio method*. The steps involved in the mole-ratio method of calculation are as follows.

Example: How many moles of HCl can be produced by letting 2 moles of H_2 react with Cl?

Step 1. Write the balanced equation, showing the reacting ratios. (Notice that we are *not* interested in how much Cl we have.)

$$H_2 + Cl_2 \longrightarrow 2HCl$$

1 mole   2 moles

Step 2. Now calculate the mole ratio by using the formula:

$$\text{mole ratio} = \frac{\text{moles of desired substance}}{\text{moles of starting substance}}.$$

$$H_2 + Cl_2 \longrightarrow 2HCl$$

1 mole of   2 moles of
starting   desired
substance   substance

$$\text{mole ratio} = \frac{2 \text{ moles HCl}}{1 \text{ mole } H_2}$$

Step 3. To find the number of moles of HCl produced, multiply the mole ratio by the number of moles of starting substance stated in the problem.

2 moles
(starting substance
from problem)

$$H_2 + Cl_2 \longrightarrow 2HCl$$

1 mole of   2 moles of
starting   desired
substance   substance

$$2 \text{ moles } H_2 \times \frac{2 \text{ moles HCl}}{1 \text{ mole } H_2} = \text{number of moles of product}$$

Cancel like units and do the calculations.

$$2 \text{ moles } H_2 \times \frac{2 \text{ moles HCl}}{1 \text{ mole } H_2} = 2 \times \frac{2 \text{ moles HCl}}{1} = 2(2 \text{ moles HCl}) = 4 \text{ moles HCl}$$

If the units do not cancel, you have set up the problem incorrectly.

4. Try the following problem.

How many moles of NH_3 can be produced by letting 9 moles of H_2 react with nitrogen?

$$3H_2 + N_2 \longrightarrow 2NH_3$$

Step 1. Write the balanced equation, showing the reacting ratios.

Step 2. Calculate the mole ratio.

Step 3. Find the number of moles of NH_3 produced.

---

5. Answers to frame 4.

Step 1. $3H_2 + N_2 \longrightarrow 2NH_3$

3 moles
of starting
substance

2 moles
of desired
substance

Step 2. mole ratio $= \dfrac{2 \text{ moles } NH_3}{3 \text{ moles } H_2}$

Step 3. 9 moles $H_2 \times \dfrac{2 \text{ moles } NH_3}{3 \text{ moles } H_2} = 9 \text{ moles } H_2 \times \dfrac{2 \text{ moles } NH_3}{3 \text{ moles } H_2} =$

$\dfrac{9(2 \text{ moles } NH_3)}{3} = 6 \text{ moles } NH_3$

---

6. Solve this mole/mole problem.

How many moles of oxygen can be produced by letting 16 moles of water decompose to produce oxygen? The balanced equation is as follows.

$$2H_2O \longrightarrow 2H_2 + O_2$$

---

7. Answers to frame 6.

Step 1. $2H_2O \longrightarrow 2H_2 + O_2$

2 moles
of starting
substance

1 mole
of desired
substance

Step 2. mole ratio $= \dfrac{1 \text{ mole } O_2}{2 \text{ moles } H_2O}$

Step 3. 16 moles $H_2O \times \dfrac{1 \text{ mole } O_2}{2 \text{ moles } H_2O} = \dfrac{16(1 \text{ mole } O_2)}{2} = 8 \text{ moles } O_2$

8. The mole/mole calculation gives answers in terms of moles. In a chemistry laboratory, however, the instruments are calibrated in grams. A simple formula allows us to
change moles to grams. It is the same formula used to solve for the number of moles
that a given weight in grams represents. The conversion formula is as follows.

$$\text{number of moles} = \frac{\text{weight in grams}}{\text{weight of 1 mole}}$$

If we multiply each side of the equation by "weight of 1 mole," we get:

$$(\text{weight of 1 mole}) \times (\text{number of moles}) = \frac{\text{weight in grams}}{\cancel{\text{weight of 1 mole}}} \times (\cancel{\text{weight of 1 mole}})$$

By canceling out "weight of 1 mole" on the right side of the equation, we are left
with the following.

$$(\text{weight of 1 mole}) \times (\text{number of moles}) = \text{weight in grams}$$

We generally turn the formula around as follows.

$$\text{weight in grams} = (\text{weight of 1 mole}) \times (\text{number of moles})$$

or, to simplify,

$$g = \frac{g}{\text{mole}} \times \text{mole}$$

9. We use the formula in the following manner.

Example: How many grams of H_2O are there in 2 moles of H_2O?

Step 1. Write the formula.

weight in grams = (weight of 1 mole) × (number of moles)

Step 2. List the data.

weight of 1 mole of H = 1.0 × 2 =   2.0 g
weight of 1 mole of O = 16.0 × 1 = 16.0 g
                                  ─────────
                                    18.0 g

number of moles = 2

Step 3. Substitute the data into the formula.

$$\text{weight in grams} = \frac{18.0 \text{ g}}{\cancel{\text{mole}}} \times 2.0 \,\cancel{\text{moles}}$$

Step 4. Do the calculation.

weight in grams = 36 g

10. How many grams are in 0.50 mole of NH_3?

Step 1. Write the formula.

Step 2. List the data.

Step 3. Substitute the data into the formula.

Step 4. Do the calculation.

11. Answers to frame 10.

Step 1. weight in grams = (weight of 1 mole) $\times$ (number of moles)

Step 2. weight of 1 mole of N = 14.0 $\times$ 1 = 14.0 g

weight of 1 mole of H = 1.0 $\times$ 3 = $\underline{3.0\text{ g}}$

17.0 g

number of moles = 0.50

Step 3. weight in grams = $\dfrac{17.0\text{ g}}{\text{mole}}$ $\times$ 0.50 ~~mole~~

Step 4. weight in grams = 8.5 g

---

12. How many grams do 2.0 moles of CO_2 weigh?

---

13. Answers to frame 12.

Step 1. weight in grams = (weight of 1 mole) $\times$ (number of moles)

Step 2. weight of 1 mole of C = 12.0 $\times$ 1 = 12.0 g

weight of 1 mole of O = 16.0 $\times$ 2 = $\underline{32.0\text{ g}}$

44.0 g

number of moles = 2.0

Step 3. weight in grams = $\dfrac{44.0\text{ g}}{\text{mole}}$ $\times$ 2.0 ~~moles~~

Step 4. weight in grams = 88 g

---

14. By combining our knowledge of how to convert moles to grams with our ability to solve mole/mole problems, we can solve mole/weight problems.

Example: How many grams of HBr are produced when 2.00 moles of H_2 react with Br?

The balanced equation is as follows.

$$H_2 + Br_2 \longrightarrow 2HBr$$

Step 1. Write the balanced equation, showing the reacting ratios.

$$H_2 + Br_2 \longrightarrow 2HBr$$

1 mole   2 moles
of starting   of desired
substance   substance

Step 2. Calculate the mole ratio.

mole ratio = $\dfrac{\text{moles desired}}{\text{moles starting}}$

mole ratio = $\dfrac{2 \text{ moles HBr}}{1 \text{ mole } H_2}$

Step 3. Calculate the moles of HBr produced.

$$2.00 \text{ moles of } H_2 \times \frac{2 \text{ moles HBr}}{1 \text{ mole } H_2} = \text{moles produced}$$

2.00 (2 moles HBr) = 4.00 moles HBr produced

Step 4. Convert the answer obtained in Step 3 from moles to grams.

weight in grams = (weight of 1 mole)(number of moles)

weight of 1 mole of H   =   1.0 × 1 =   1.0 g
weight of 1 mole of Br = 79.9 × 1 = 79.9 g
                                                        ─────
                                                        80.9 g

number of moles = 4.00 moles

$$\text{weight in grams} = \frac{80.9 \text{ g}}{\text{mole}} \times 4.00 \text{ moles}$$

weight in grams = 324 g

Solve this problem.

How many grams of ammonia will 6.00 moles of hydrogen reacting with nitrogen produce? The balanced equation is as follows.

$$3H_2 + N_2 \longrightarrow 2NH_3$$

Step 1. Write the equation, showing the reacting ratios.

Step 2. Write the mole ratio.

Step 3. Calculate how many moles of the desired substance were produced.

Step 4. Change moles to grams.

15. Answers to frame 14.

Step 1. $3H_2 + N_2 \longrightarrow 2NH_3$
           3 moles   2 moles

Step 2.   mole ratio $= \dfrac{2 \text{ moles } NH_3}{3 \text{ moles } H_2}$

Step 3. $6.00 \; \text{moles } H_2 \times \dfrac{2 \text{ moles } NH_3}{3 \text{ moles } H_2} = $ moles produced

           $6.00 \; \dfrac{(2 \text{ moles } NH_3)}{3} = 4.00 \text{ moles } NH_3$

Step 4.   weight in grams $=$ (weight of 1 mole)(number of moles)
           weight of 1 mole of N $= 14.0 \times 1 = 14.0$ g
           weight of 1 mole of H $= 1.0 \times 3 = \underline{3.0}$ g
           17.0 g

           number of moles $= 4.00$ moles

           weight in grams $= \dfrac{17.0 \text{ g}}{\text{mole}} \times 4.00 \; \text{moles}$

           weight in grams $= 68.0$ g

16. Which formula below is the correct formula for converting moles to grams? _____

    a.   weight in grams $=$ (weight of 1 mole)(number of moles)

    b.   weight in grams $= \dfrac{\text{weight in grams}}{\text{weight of 1 mole}}$

17. Solve the following problems.   16. a

    a. How many moles of O_2 can be produced by letting 8 moles of HgO decompose? The balanced equation is as follows.

    $$2HgO \longrightarrow 2Hg + O_2$$

    b. How many moles of Fe_2O_3 can be produced by letting 9 moles of Fe react with O_2? The balanced equation is as follows.

    $$4Fe + 3O_2 \longrightarrow 2Fe_2O_3$$

    c. How many moles of O_2 will be obtained by decomposing 3.50 moles of $KClO_3$? The balanced equation is as follows.

    $$2KClO_3 \longrightarrow 2KCl + 3O_2$$

    d. How many moles of NaCl can be made from 1.41 moles of Cl_2 in the following reaction?

    $$2Na + Cl_2 \longrightarrow 2NaCl$$

Convert the following moles of compounds to grams.

e. 4.50 moles of $Ca(OH)_2$

f. 2.5 moles of CO

Solve the following mole/weight problems.

g. 10.0 moles of H_2 will produce how many grams of H_2O? The balanced equation is as follows.

$$2H_2 + O_2 \longrightarrow 2H_2O$$

h. 5.00 moles of C will produce how many grams of CO? The balanced equation is as follows.

$$2C + O_2 \longrightarrow 2CO$$

i. How many grams of HNO_3 will be produced from 2.50 moles of N_2O_5 in the following reaction?

$$N_2O_5 + H_2O \longrightarrow 2HNO_3$$

j. How many grams of Mg_3N_2 will be produced from 0.750 moles of Mg in the following reaction?

$$3Mg + N_2 \longrightarrow Mg_3N_2$$

---

18. Answers to frame 17.

a.   8 moles HgO $\times \dfrac{1 \text{ mole } O_2}{2 \text{ moles HgO}}$ = 4 moles O_2

b.   9 moles Fe $\times \dfrac{2 \text{ moles } Fe_2O_3}{4 \text{ moles Fe}}$ = 4.5 moles Fe_2O_3

c.   3.50 moles $KClO_3 \times \dfrac{3 \text{ moles } O_2}{2 \text{ moles } KClO_3}$ = 5.25 moles O_2

d.   1.41 moles $Cl_2 \times \dfrac{2 \text{ moles NaCl}}{1 \text{ mole } Cl_2}$ = 2.82 moles NaCl

e. $4.50 \text{ moles Ca(OH)}_2 \times \dfrac{74.1 \text{ g Ca(OH)}_2}{1 \text{ mole Ca(OH)}_2} = 333 \text{ g Ca(OH)}_2$

f. $2.5 \text{ moles CO} \times \dfrac{28.0 \text{ g CO}}{1 \text{ mole CO}} = 70 \text{ g CO}$

g. $10.0 \text{ moles H}_2 \times \dfrac{2 \text{ moles H}_2\text{O}}{2 \text{ moles H}_2} \times \dfrac{18.0 \text{ g H}_2\text{O}}{1 \text{ mole H}_2\text{O}} = 180 \text{ g H}_2\text{O}$

h. $5.00 \text{ moles C} \times \dfrac{2 \text{ moles CO}}{2 \text{ moles C}} \times \dfrac{28.0 \text{ g CO}}{1 \text{ mole CO}} = 140 \text{ g CO}$

i. $2.50 \text{ moles N}_2\text{O}_5 \times \dfrac{2 \text{ moles HNO}_3}{1 \text{ mole N}_2\text{O}_5} \times \dfrac{63.0 \text{g HNO}_3}{1 \text{ mole HNO}_3} = 315 \text{ g HNO}_3$

j. $0.750 \text{ moles Mg} \times \dfrac{1 \text{ mole Mg}_3\text{N}_2}{3 \text{ moles Mg}} \times \dfrac{100.9 \text{g Mg}_3\text{N}_2}{1 \text{ mole Mg}_3\text{N}_2} = 25.2 \text{g Mg}_3\text{N}_2$

Part 2. More-Complicated Stoichiometry Calculations

1. Given the weight of one reactant and the balanced equation, the amount of the second reactant needed can be calculated.

Example: How many moles of NaOH are needed to react with 4.5 moles of H_2SO_4? The balanced equation is as follows.

$$H_2SO_4 + 2NaOH \longrightarrow Na_2SO_4 + 2H_2O$$

Step 1. $H_2SO_4 + 2NaOH \longrightarrow Na_2SO_4 + 2H_2O$

1 mole   2 moles
of   of
starting   desired
substance   substance

Step 2. $\text{mole ratio} = \dfrac{2 \text{ moles NaOH} \ (\text{desired})}{1 \text{ mole H}_2\text{SO}_4 \ (\text{starting})}$

Step 3. $4.5 \ \cancel{\text{moles H}_2\text{SO}_4} \times \dfrac{2 \text{ moles NaOH}}{1 \ \cancel{\text{mole H}_2\text{SO}_4}} = 9.0 \text{ moles NaOH}$

(stated in problem)

Try these problems.

a. How many moles of Cr are needed to react with 8.00 moles of O_2 in the following equation?

$$4Cr + 3O_2 \longrightarrow 2Cr_2O_3$$

b. How many moles of HNO_3 are needed to react with 0.50 moles of $CaCl_2$ in the following equation?

$$CaCl_2 + 2HNO_3 \longrightarrow Ca(NO_3)_2 + 2HCl$$

2. Answers to frame 1.

   a.   Step 1. Cr: 4 moles of desired substance
                  O_2:   3 moles of starting substance

         Step 2. mole ratio $= \dfrac{4 \text{ moles Cr}}{3 \text{ moles } O_2}$

         Step 3. 8.00 moles O_2 $\times$ $\dfrac{4 \text{ moles Cr}}{3 \text{ moles } O_2} = 10.7$ moles Cr

   b.   Step 1. HNO_3 = 2 moles desired substance
                  $CaCl_2$ = 1 mole starting substance

         Step 2. mole ratio $= \dfrac{2 \text{ moles } HNO_3}{1 \text{ mole } CaCl_2}$

         Step 3. 0.50 mole $CaCl_2$ $\times$ $\dfrac{2 \text{ moles } HNO_3}{1 \text{ mole } CaCl_2} = 1.0$ mole HNO_3

---

3. To solve a weight/weight problem, you start with grams, convert to moles to solve a mole/mole problem, and then convert your answer back to grams. The sequence is as follows.

Step 1.   Convert grams of starting substance to moles of starting substance.

Step 2.   Calculate the mole ratio.

Step 3.   Multiply the moles of starting substance (from Step 1) by the mole ratio to find the moles of desired substance.

Step 4.   Change the moles of desired substance to grams.

Example: What weight of CO_2 is produced from 90.0 grams of C_2H_6? The balanced equation is:

$$2C_2H_6 + 7O_2 \longrightarrow 4CO_2 + 6H_2O$$

Step 1.   Convert grams of starting substance to moles.

    a.   number of moles $= \dfrac{\text{wt in g}}{\text{wt/mole}}$

    b.   wt in g = 90.0 g
         wt/mole = 30.0 g/mole

    c.   number of moles $= \dfrac{90.0 \text{ g}}{30.0 \text{ g/mole}}$

    d.   number of moles = 3.00 moles of C_2H_6 (needed in Step 3)

Step 2.   Calculate the mole ratio.

    mole ratio $= \dfrac{\text{moles desired}}{\text{moles starting}}$

    mole ratio $= \dfrac{4.00 \text{ moles } CO_2}{2.00 \text{ moles } C_2H_6}$

Step 3.   Find the moles of desired substance using the moles of starting substance calculated in Step 1.

    3.00 moles C_2H_6 $\times$ $\dfrac{4 \text{ moles } CO_2}{2 \text{ moles } C_2H_6}$ $= 6.00$ moles CO_2

Step 4. Change the moles of desired substance (CO_2) to grams.

    a. wt in g = (number of moles) (wt/mole)

    b. number of moles = 6.00 moles
       wt/mole = 44.0 g/mole

    c. wt in g = 6.00 ~~moles~~ $\times \dfrac{44.0\ g}{\text{mole}}$

    d. wt in g = 264 g of CO_2

An alternate method of solving this problem uses a series of conversion factors. This method is faster, but make sure you understand all the steps involved before using it.

$$90.0\ \text{g}\ \cancel{C_2H_6} \times \frac{1\ \text{mole}\ \cancel{C_2H_6}}{30.0\ \text{g}\ \cancel{C_2H_6}} \times \frac{4.00\ \text{mole}\ \cancel{CO_2}}{2.00\ \text{mole}\ \cancel{C_2H_6}} \times \frac{44.0\ \text{g}\ CO_2}{1\ \text{mole}\ \cancel{CO_2}} = 264\ \text{g}\ CO_2$$

                   ↑   ↑   ↑

        converts g   uses mole ratio   converts
        to moles   to solve for   moles to g
                        moles of desired
                          substance

Try these problems.

a. What weight of HNO_3 can be produced from 138 g of NO_2?
The balanced equation is:

$$3NO_2 + H_2O \longrightarrow 2HNO_3 + NO$$

b. What weight of NH_3 is needed to react with 196.4 g of H_2SO_4? The balanced equation is:

$$2NH_3 + H_2SO_4 \longrightarrow (NH_4)_2SO_4$$

---

4. Answers to frame 3.

  a. $138\ \text{g}\ NO_2 \times \dfrac{1\ \text{mole}\ NO_2}{46.0\ \text{g}\ NO_2} \times \dfrac{2\ \text{moles}\ HNO_3}{3\ \text{moles}\ NO_2} \times \dfrac{63.0\ \text{g}\ HNO_3}{1\ \text{mole}\ HNO_3} = 126\ \text{g}\ HNO_3$

  b. $196.4\ \text{g}\ H_2SO_4 \times \dfrac{1\ \text{mole}\ H_2SO_4}{98.1\ \text{g}\ H_2SO_4} \times \dfrac{2\ \text{moles}\ NH_3}{1\ \text{mole}\ H_2SO_4} \times \dfrac{17.0\ \text{g}\ NH_3}{1\ \text{mole}\ NH_3} = 68.1\ \text{g}\ NH_3$

Part 3. Limiting Reagent

1. The weight of one or more reactants can be in excess of the exact ratio indicated by the equation. The reactant that is *not* in excess is called the limiting reagent. The limiting reagent will be used up first during the reaction, and, when this reactant is gone, no more product can form. Limiting-reagent problems can be identified by noting that the amounts of two reactants are given.

Example: How many grams of $MgCl_2$ can be produced from 16.14 g of $Mg(OH)_2$ and 10.97 g of HCl? The balanced equation is:

$$Mg(OH)_2 \; + \; 2HCl \longrightarrow \; MgCl_2 \; + \; 2H_2O$$

Step 1.   To determine which reactant ($Mg(OH)_2$ or HCl) is the limiting reagent, calculate the number of moles of each reactant.

$$\text{moles of } Mg(OH)_2 \; = \frac{\text{wt in g}}{\text{wt/mole}}$$

$$\text{moles of } Mg(OH)_2 \; = \frac{16.14 \text{ g}}{58.3 \text{ g/mole}} = 0.277 \text{ mole}$$

$$\text{moles of HCl} \; = \frac{\text{wt in g}}{\text{wt/mole}}$$

$$\text{moles of HCl} \; = \frac{10.97 \text{ g}}{36.5 \text{ g/mole}} = 0.301 \text{ mole}$$

Step 2.   Determine the limiting reagent by finding the ratio of moles calculated in Step 1 and comparing this ratio to the mole ratio in the equation.

$$\text{Ratio of moles calculated} \; = \frac{0.277/0.277}{0.301/0.277}$$

$$= \frac{1.00 \text{ mole } Mg(OH)_2}{1.09 \text{ moles HCl}}$$

$$\text{Mole ratio from equation} \; = \frac{1 \text{ mole } Mg(OH)_2}{2 \text{ mole HCl}}$$

Two moles of HCl are needed for every 1 mole of $Mg(OH)_2$. We have only 1.09 moles of HCl per 1 mole of $Mg(OH)_2$. HCl is the limiting reagent and is used in the calculation.

Step 3.   Use the original number of moles of HCl calculated in Step 1 to solve for the number of grams of $MgCl_2$ produced.

$$0.301 \text{ moles HCl} \times \frac{1 \text{ mole } MgCl_2}{2 \text{ moles HCl}} = 0.151 \text{ moles } MgCl_2$$

$$\text{wt in g of } MgCl_2 \; = \; (\text{number of moles})(\text{wt/mole})$$

$$\text{wt in g of } MgCl_2 \; = \; 0.151 \text{ moles } MgCl_2 \times \frac{95.3 \text{ g } MgCl_2}{1 \text{ mole } MgCl_2}$$

$$\text{wt in g} \; = \; 14.4 \text{ g } MgCl_2$$

To shorten the calculation, this last step could be written as a series of conversion factors.

$$0.301 \text{ moles HCl} \times \frac{1 \text{ mole } MgCl_2}{2 \text{ moles HCl}} \times \frac{95.3 \text{ g } MgCl_2}{1 \text{ mole } MgCl_2} = 14.4 \text{ g } MgCl_2$$

Try these problems:

a. How many grams of NaCl can be produced from 50.0 grams of Na and 100 grams of Cl_2? The balanced equation is:

$$2Na + Cl_2 \longrightarrow 2NaCl$$

b. How many grams of AgI can be produced from 50.0 grams of CaI_2 and 20.0 grams of $AgNO_3$? The balanced equation is:

$$CaI_2 + AgNO_3 \longrightarrow Ca(NO_3)_2 + 2AgI$$

2. Answers to frame 1.

a. Step 1. moles of Na $= \dfrac{50.0 \text{ g}}{23.0 \text{ g/mole}} = 2.17$

moles of Cl_2 $= \dfrac{100 \text{ g}}{71.0 \text{ g/mole}} = 1.41$

Step 2. ratio of moles calculated $= \dfrac{2.17/1.41}{1.41/1.41} = \dfrac{1.54 \text{ moles Na}}{1.00 \text{ mole } Cl_2}$

mole ratio from equation $= \dfrac{2 \text{ moles Na}}{1 \text{ mole } Cl_2}$

Na is the limiting reagent and is used in the calculation.

Step 3. 2.17 moles Na $\times \dfrac{2 \text{ moles NaCl}}{2 \text{ moles Na}} \times \dfrac{58.5 \text{ g NaCl}}{1 \text{ mole NaCl}} = 127 \text{ g NaCl}$

b. Step 1. moles CaI_2 $= \dfrac{50.0 \text{ g}}{293.9 \text{ g/mole}} = 0.170$

moles $AgNO_3$ $= \dfrac{20.0 \text{ g}}{169.9 \text{ g/mole}} = 0.118$

Step 2. ratio of moles calculated $= \dfrac{0.170/0.118}{0.118/0.118} = \dfrac{1.44 \text{ moles } CaI_2}{1.00 \text{ mole } AgNO_3}$

mole ratio from equation $= \dfrac{1.00 \text{ mole } CaI_2}{1.00 \text{ mole } AgNO_3}$

CaI_2 is in excess, so $AgNO_3$ is the limiting reagent.

Step 3. 0.118 mole $AgNO_3$ $\times \dfrac{2 \text{ mole AgI}}{1 \text{ mole } AgNO_3} \times \dfrac{234.8 \text{ g AgI}}{1 \text{ mole AgI}} = 55.4 \text{ g AgI}$

3. Solve the following problems.

   a. How many moles of Al are needed ro react with 4.5 moles of Br_2? The balanced equation is as follows.

$$2Al + 3Br_2 \longrightarrow 2AlBr_3$$

   b. How many moles of O_2 are needed to react with 2.5 moles of ethanol, C_2H_5OH?

$$2C_2H_5OH + 6O_2 \longrightarrow 4CO_2 + 6H_2O$$

   c. What weight of HCN is produced from 30.0 grams of CH_4?

$$2CH_4 + 3O_2 + 2NH_3 \longrightarrow 2HCN + 6H_2O$$

   d. How many grams of NaCl can be produced from 8.50 g of Na?

$$2Na + Cl_2 \longrightarrow 2NaCl$$

   e. How many grams of Al_2O_3 are produced from 150 grams of Fe_2O_3 and 150 grams of Al?

$$Fe_2O_3 + 2Al \longrightarrow Al_2O_3 + 2Fe$$

   f. How many grams of CaC_2 are produced from 200 g of CaO and 100 g of C?

$$CaO + 3C \longrightarrow CaC_2 + CO$$

4. Answers to frame 3.

a. $4.5 \text{ moles } Br_2 \times \dfrac{2 \text{ moles Al}}{3 \text{ moles } Br_2} = 3.0 \text{ moles Al}$

b. $2.5 \text{ moles } C_2H_5OH \times \dfrac{6 \text{ moles } O_2}{2 \text{ moles } C_2H_5OH} = 7.5 \text{ moles } O_2$

c. $30.0 \text{ g } CH_4 \times \dfrac{1 \text{ mole } CH_4}{16.0 \text{ g } CH_4} \times \dfrac{2 \text{ moles HCN}}{2 \text{ moles } CH_4} \times \dfrac{27.0 \text{ g HCN}}{1 \text{ mole HCN}} = 50.6 \text{ g HCN}$

d. $8.50 \text{ g Na} \times \dfrac{1 \text{ mole Na}}{23.0 \text{ g Na}} \times \dfrac{2 \text{ moles NaCl}}{2 \text{ moles Na}} \times \dfrac{58.5 \text{ g NaCl}}{1 \text{ mole NaCl}} = 21.6 \text{ g NaCl}$

e. Step 1. $\text{moles } Fe_2O_3 = \dfrac{150 \text{ g}}{159.6 \text{ g/mole}} = 0.940$

$\text{moles Al} = \dfrac{150 \text{ g}}{27.0 \text{ g/mole}} = 5.56$

Step 2. $\text{ratio of moles calculated} = \dfrac{0.940/0.940}{5.56/0.940} = \dfrac{1.00 \text{ mole } Fe_2O_3}{5.91 \text{ moles Al}}$

$\text{mole ratio from equation} = \dfrac{1.00 \text{ mole } Fe_2O_3}{2.00 \text{ moles Al}}$

Al is in excess, so Fe_2O_3 is the limiting reagent.

Step 3. $0.940 \text{ mole } Fe_2O_3 \times \dfrac{1 \text{ mole } Al_2O_3}{1 \text{ mole } Fe_2O_3} \times \dfrac{102 \text{ g } Al_2O_3}{1 \text{ mole } Al_2O_3} = 95.9 \text{ g } Al_2O_3$

f. Step 1. $\text{moles CaO} = \dfrac{200 \text{ g}}{56.1 \text{ g/mole}} = 3.57$

$\text{moles C} = \dfrac{100 \text{ g}}{12.0 \text{ g/mole}} = 8.33$

Step 2. $\text{ratio of moles calculated} = \dfrac{3.57/3.57}{8.33/3.57} = \dfrac{1.00 \text{ mole CaO}}{2.33 \text{ mole C}}$

$\text{mole ratio from equation} = \dfrac{1 \text{ mole CaO}}{3 \text{ mole C}}$

C is the limiting reagent.

Step 3. $8.33 \text{ mole C} \times \dfrac{1 \text{ mole } CaC_2}{3 \text{ mole C}} \times \dfrac{64.1 \text{ g } CaC_2}{1 \text{ mole } CaC_2} = 178 \text{ g } CaC_2$

CHEMICAL EQUATIONS AND STOICHIOMETRY
EVALUATION TEST 1
PART 1

1. In the blanks below each formula, indicate how many moles are represented by the coefficients.

$$2C_4H_{10} \quad + \quad 13O_2 \quad \longrightarrow \quad 8CO_2 \quad + \quad 10H_2O$$

_____ _____ _____ _____

2. List the steps needed to solve a mole/mole problem.

a. _____

b. _____

c. _____

Solve the following mole/mole problem.

3. How many moles of O_2 can be produced by letting 12.0 moles of $KClO_3$ react? The balanced equation is as follows.

$$2KClO_3 \quad \longrightarrow \quad 2KCl \quad + \quad 3O_2$$

4. Write the formula for changing grams to moles.

5. Write the formula for changing moles to grams.

6. 2.00 moles of $Ca(OH)_2$ weigh how many grams?

Solve the following mole/weight problem.

7. 12.0 moles of $NaClO_3$ will produce how many grams of O_2? The balanced equation is as follows.

$$2NaClO_3 \quad \longrightarrow \quad 2NaCl \quad + \quad 3O_2$$

CHEMICAL EQUATIONS AND STOICHIOMETRY
EVALUATION TEST 2
PART 2-PART 3

1. How many moles of Cu are needed to react with 3.5 moles of $AgNO_3$?

$$Cu \quad + \quad 2AgNO_3 \quad \longrightarrow \quad Cu(NO_3)_2 \quad + \quad 2Ag$$

2. List the steps needed to solve a weight/weight problem.

a. _____

b. _____

c. _____

d. _____

3. What weight of NaOH is produced from 120 grams of Na_2O?

$$Na_2O \quad + \quad H_2O \quad \longrightarrow \quad 2NaOH$$

4. What weight of KCl is produced from 250 g of K and 100 g of Cl_2?

$$2K \quad + \quad Cl_2 \quad \longrightarrow \quad 2KCl$$

UNIT 13
The Concentration of Solutions

In this unit you will learn some basic concepts about solutions. The definition of a solution, the components of a solution, and the common types of solutions will be discussed. You will learn the most common methods of measuring the concentration of solutions and you will learn how to use dilution formulas. You will learn how to make calculations using the concepts of normality *and* equivalent weight. *You will make a titration calculation and find the pH of a solution.*

Part 1. Definition of a Solution

1. In the unit on the divisions and properties of matter (Unit 7), you learned that matter can be divided into pure substances and mixtures. Pure substances contain all one kind of particle. Mixtures are made up of two or more pure substances, physically combined. *Physical combination* means that the particles of matter in the mixture do not alter their identity; a water molecule is still a water molecule even if some sugar is put in the water. Solutions are usually mixtures.

   Matter can be divided into the two subdivisions of (a) _____

   _____ and (b) _____ .

2. Which of the following describes the composition of a mixture? _____

   a. all one kind of atom
   b. several pure substances
   c. all one kind of molecule

3. Are solutions usually compounds or mixtures? _____

1. a. pure substances
   b. mixtures

2. b

Part 2. Components of a Solution

1. A *solution* is a mixture composed of two or more soluble pure substances, physically combined. (A *soluble* substance is one that will dissolve in another substance.) In this unit, we will deal only with solutions composed of two pure substances.

   We give the names *solute* and *solvent* to the two pure substances that compose a solution. The *solute* is the substance that is dissolved and the *solvent* is the substance that does the dissolving. When solids are dissolved in liquids, the solid is the solute and the liquid is the solvent.

   Example: When salt is dissolved in water, the salt is the solute and the water is the solvent.

   Which of the following is the correct name for the substance that is dissolved in a solution? _____

   a.   solution   b.   solvent   c.   solute

2. Which of the following is the correct name for the substance that does the dissolving in a solution? _____

   a.   solution   b.   solvent   c.   solute

3. In a sugar and water solution, the water would be called the _____ .

4. When liquids are dissolved in liquids, one of the liquids is usually water. The water is usually considered the solvent and the other liquid the solute.

   Example: In solutions of alcohol and water, the alcohol is the solute and the water is the solvent.

   In a liquid/liquid solution, water would be considered the _____ .

   a.   solvent   b.   solution   c.   solute

5. Antifreeze, when dissolved in water, is considered the _____ .

Part 3. Concentrations of Solutions

1. The concentration of a solution can be expressed in terms of how much solute is dissolved in a given volume of solution. The terms *dilute* and *concentrated* are often used to indicate the amount of solute in a given volume of solution. These terms are imprecise; that is, they describe general quantities, rather than specific amounts. *Dilute* means that there is a small amount of solute in a unit of solution. *Concentrated* means that there is a large amount of solute in a unit of solution.

   Concentrations of solutions are usually expressed in terms of how much (a) _____ _____ is dissolved in a given volume of (b) _____ .

2. A solution containing a large amount of solute is called a _____ solution.

---

3. mixtures

1. c

2. b

3. solvent

4. solvent

5. solute

1. a. solute
   b. solution

3. A small amount of solute in a given volume of solution is called a _____ solution.

2. concentrated

4. Two other terms used to indicate the concentration of a solution are *saturated* and *unsaturated*. A saturated solution contains all of the dissolved solute it can at a given temperature. Only a certain amount of sugar, for example, can be dissolved in a cup of coffee. At a certain point, any additional sugar will not be dissolved and will sit at the bottom of the cup. At this point, we say the solution is saturated. An unsaturated solution is able to dissolve more solute than is present in the solution.

   A solution that has dissolved all of the solute it can at a given temperature is said to be _____ .

3. dilute

5. An unsaturated solution can/cannot dissolve more solute.

4. saturated

6. A solution that contains a large amount of solute in a given volume of solution is called _____ .

   a. concentrated   b. dilute   c. saturated   d. unsaturated

5. can

7. A solution that can dissolve more solute that it already has is called _____ .

   a. concentrated   b. dilute   c. saturated   d. unsaturated

6. a. concentrated

8. A solution that contains a small amount of solute in a given volume of solution is called _____ .

   a. concentrated   b. dilute   c. saturated   d. unsaturated

7. d. unsaturated

9. A solution that has dissolved all of the solute it can at a given temperature is called _____ .

   a. concentrated   b. dilute   c. saturated   d. unsaturated

8. dilute

10. The most precise method of expressing the concentration of solutions is to state exactly what weight or volume of solute is dissolved in an exact amount of solution. This is done in several ways.

    The first method we shall look at is called *percentage by weight*. This method gives the weight of solute dissolved in a definite weight of solution and is usually expressed in the following mathematical-equation form.

$$\text{percentage by weight} = \frac{\text{grams of solute}}{\text{grams of solution}} \times 100\%$$

    To solve a percentage-by-weight problem, you must know the number of grams of solute dissolved and the number of grams of solution. The answer is multiplied by 100% to put it in percent form.

    In order to express the concentration of a solution in percentage by weight, we must know what weight of solute is dissolved in what weight of _____ .

9. saturated

11. Is it necessary to know the volume of a solution in order to solve a percentage-by-weight problem? _____

10. solution

12. The formula for finding the percentage-by-weight concentration of a solution is as follows.

    percentage by weight = _____

11. no

13. Determining the concentration of a solution by using the percentage-by-weight formula involves four steps.

    Example: 8.0 g of salt are dissolved in 40.0 g of solution. What is the percentage by weight of this solution?

    The problem is solved in the following steps.

    Step 1. List the formula.

    $$\text{percentage by weight} = \frac{\text{g solute}}{\text{g solution}} \times 100\%$$

    Step 2. List the data.

    g solute = 8.0 g
    g solution = 40.0 g

    Step 3. Substitute the data into the formula.

    $$\text{percentage by weight} = \frac{8.0 \text{ g}}{40.0 \text{ g}} \times 100\%$$

    Step 4. Do the calculation.

    percentage by weight = 20%

    Determine the concentration of this solution in percentage by weight.

                    20 g of sugar dissolved in 80 g of solution

    Step 1.

    Step 2.

    Step 3.

    Step 4.

12. $\dfrac{\text{g solute}}{\text{g solution}} \times 100\%$

14. Answers to frame 13.

    Step 1.   percentage by weight $= \dfrac{\text{g solute}}{\text{g solution}} \times 100\%$

    Step 2.   grams solute = 20 g
              grams solution = 80 g

    Step 3.   percentage by weight $= \dfrac{20 \text{ g}}{80 \text{ g}} \times 100\%$

    Step 4.   percentage by weight = 25%

15. Usually the weight of the solution is not given; the weight of the solute and the weight of the solvent are given instead. The weight of the solution can be calculated simply by adding these up.

> *Example:*   10 g of salt are dissolved in 30 g of water. What is the percentage by weight of this solution?

Step 1. List the formula.

$$\text{percentage by weight} = \frac{\text{g solute}}{\text{g solution}} \times 100\%$$

Step 2. List the data.

g solute = 10 g
g solution = (g solute + g solvent) = 10 g + 30 g = 40 g

Step 3. Substitute the data into the formula.

$$\text{percentage by weight} = \frac{10 \text{ g}}{40 \text{ g}} \times 100\%$$

Step 4. Do the calculation.

percentage by weight = 25%

The weight of the solution is found by adding the weight of (a) _____ to the weight of (b) _____ .

16. The formula for finding the weight of a solution, given the weight of solute and solvent, is as follows.

weight of solution = _____

17. Answers to frame 16.

weight of solute + weight of solvent or g solute + g solvent

18. In a solution of 20 g of salt and 40 g of water,

a.   g solute = _____,

b.   g solvent = _____,

c.   g solution = _____ .

19. Solve the following percentage-by-weight problem.

> 30 g of salt dissolved in 30 g of water

What is the percentage by weight of this solution?

Step 1.

Step 2.

Step 3.

Step 4.

15. a. solute
    b. solvent

18. a. 20 g
    b. 40 g
    c. 60 g

20. Answers to frame 19.

Step 1.   percentage by weight $= \dfrac{\text{g solute}}{\text{g solution}} \times 100\%$

Step 2.   g solute $= 30$ g
g solution $= 30$ g $+ 30$ g $= 60$ g

Step 3.   percentage by weight $= \dfrac{30 \text{ g}}{60 \text{ g}} \times 100\%$

Step 4.   percentage by weight $= 50\%$

21. Solve this problem.

10 g of sugar dissolved in 90 g of water

What is the percentage by weight of this solution?

Step 1.

Step 2.

Step 3.

Step 4.

22. Another method of expressing the concentration of a solution is called *percentage by volume*. This method gives the volume of liquid solute dissolved in a definite volume of solution. The mathematical expression for percentage by volume is as follows.

percentage by volume $= \dfrac{\text{volume of liquid solute}}{\text{volume of solution}} \times 100\%$

To solve a percentage-by-volume problem, you must know the volume of liquid solute dissolved in the stated volume of solution. The answer is multiplied by 100% to put it in percentage form.

To determine the concentration of a solution in percentage by volume, we must know what volume of solute is dissolved in a given volume of _____ .

21. 10%

23. Is it necessary to know the weight of the solute in order to solve a percentage-by-volume problem? _____

22. solution

24. Which of the formulas below is used to determine the percentage-by-volume concentration of a solution? _____

a. $\dfrac{\text{g solute}}{\text{g solution}} \times 100\%$

b. $\dfrac{\text{volume of solute}}{\text{volume of solution}} \times 100\%$

23. no

25. The percentage-by-volume formula is used in the following manner.   24. b

    Example: 40 mL of alcohol added to enough water to make 100 mL of solution.
               What is the percentage by volume of this solution?

    Step 1. Write the formula.

    $$\text{percentage by volume} = \frac{\text{volume of solute}}{\text{volume of solution}} \times 100\%$$

    Step 2. List the data.

    volume of solute = 40 mL
    volume of solution = 100 mL

    Step 3. Substitute the data into the formula.

    $$\text{percentage by volume} = \frac{40 \text{ mL}}{100 \text{ mL}} \times 100\%$$

    Step 4. Do the calculation.

    $$\text{percentage by volume} = 40\%$$

    Solve the following percentage-by-volume problem.

    2.0 L of antifreeze are added to enough water to make 6.0 L of solution. What is the percentage by volume of this solution?

    Step 1.

    Step 2.

    Step 3.

    Step 4.

---

26. Answers to frame 25.

    Step 1. $\text{percentage by volume} = \dfrac{\text{volume of solute}}{\text{volume of solution}} \times 100\%$

    Step 2. volume of solute = 2.0 L
             volume of solution = 6.0 L

    Step 3. $\text{percentage by volume} = \dfrac{2.0\ L}{6.0\ L} \times 100\%$

    Step 4. percentage by volume = 33%

27. A very useful method of measuring the concentration of a solution is called *molarity*. Molarity is used so often in chemistry that we abbreviate the word as *M*.

Molarity expresses the number of moles of solute dissolved in a given volume of solution. The volume is always expressed in liters. Molarity is expressed in equation form as follows.

$$\text{Molarity } (M) = \frac{\text{moles of solute}}{\text{liters of solution}}$$

To determine the concentration of a solution, expressed in molarity, we need to know how many (a) _____ of solute are dissolved in how many (b) _____ of solution.

28. In molarity problems the volume of the solution must always be expressed in _____ _____ .

27. a. moles
    b. liters

29. Which formula below is used to solve for the molarity of a solution? _____

   a. $\dfrac{\text{volume of solute}}{\text{volume of solution}} \times 100\%$ b. $\dfrac{\text{grams of solute}}{\text{grams of solution}} \times 100\%$

   c. $\dfrac{\text{moles of solute}}{\text{liters of solution}}$

28. liters

30. If you were given the number of grams of solute and the number of grams of solvent in a solution, which of the following could you solve for? _____

   a. molarity   b. percentage by weight   c. percentage by volume

29. c

31. The abbreviation for molarity is _____ .

30. b

32. The molarity formula is used as follows.

   Example: 6 moles of salt dissolved in 2 *L* of solution. What is the molarity of this solution?

Step 1. Write the formula.

$$M = \frac{\text{moles of solute}}{\text{liters of solution}}$$

Step 2. List the data.

   moles of solute = 6 moles
   liters of solution = 2 *L*

Step 3. Substitute the data into the formula.

$$M = \frac{6 \text{ moles}}{2 \text{ } L}$$

Step 4. Do the calculation.

$$M = 3 \text{ moles}/L$$

31. *M*

Solve the following molarity problem.

2 moles of sugar are dissolved in 4 L of solution. What is the molarity of this solution?

Step 1.

Step 2.

Step 3.

Step 4.

---

33. Answers to frame 32.

Step 1. $M = \dfrac{\text{moles of solute}}{\text{liters of solution}}$

Step 3. $M = \dfrac{2 \text{ moles}}{4\ L}$

Step 2. moles of solute = 2 moles
liters of solution = 4 L

Step 4. $M = 0.5$ mole/L

---

34. The number of moles of solute is not always given outright. Frequently the weight of the solute is given in grams. You already know that to convert grams to moles you use the following formula.

$$\text{number of moles} = \frac{\text{weight in grams}}{\text{weight of 1 mole}}$$

The following problem illustrates the situation in which the weight of the solute is given in grams.

Example: 80 g of NaOH are dissolved in 4 L of solution. What is the molarity of this solution?

Step 1. Write the formula.

$$M = \frac{\text{moles of solute}}{\text{liters of solution}}$$

Step 2. List the data.

moles of solute = ?

This answer is not given directly in the example, so we must use the conversion formula.

$$\text{number of moles} = \frac{\text{weight in grams}}{\text{weight of 1 mole}}$$

At this point we have a problem within a problem. We go through the same four steps to solve this problem that we are using to solve the original problem. We already have the formula.

Step 1a. moles $= \dfrac{\text{weight in grams}}{\text{weight of 1 mole}}$

Step 2a. weight in grams = 80 g

weight of 1 mole of Na = 23.0 × 1 = 23.0 g
weight of 1 mole of O = 16.0 × 1 = 16.0 g
weight of 1 mole of H = 1.0 × 1 = 1.0 g
                                    ⎯⎯⎯⎯⎯
                                    40.0 g

Step 3a.   number of moles $= \dfrac{80 \text{ g}}{40.0 \text{ g}}$

Step 4a.   number of moles $= 2.0$

We can now take this answer and use it in Step 2 above of the original problem.

Step 2.   (continued)

     moles of solute $= 2.0$ moles
     liters of solution $= 4\ L$

Step 3.   Substitute the data into the formula.

$$M = \dfrac{2.0 \text{ moles}}{4\ L}$$

Step 4.   Do the calculation.

$$M = 0.5 \text{ mole}/L$$

Here is another example of this type of problem in a more condensed form.

Example: 68 g of $CaSO_4$ are dissolved in 4.0 L of solution. What is the molarity of this solution?

Step 1. $M = \dfrac{\text{moles of solute}}{\text{liters of solution}}$

Step 2.   moles of solute $= \dfrac{\text{weight in grams}}{\text{weight of 1 mole}}$

     weight in grams $= 68$ g
     weight of 1 mole of Ca $= 40.1 \times 1 = 40.1$ g
     weight of 1 mole of S $\ = 32.1 \times 1 = 32.1$ g
     weight of 1 mole of O $\ = 16.0 \times 4 = \underline{64.0}$ g
                                       136.2 g

     moles of solute $= \dfrac{68 \text{ g}}{136.2 \text{ g}}$

     moles of solute $= 0.50$ mole
     liters of solution $= 4.0\ L$

Step 3. $M = \dfrac{0.50 \text{ mole}}{4.0\ L}$

Step 4. $M = 0.13 \text{ mole}/L$

Solve the following problem.

71 g of Na_2SO_4 are dissolved in 6.0 L of solution. What is the molarity of this solution?

35. Answers to frame 34.

Step 1. $M = \dfrac{\text{moles of solute}}{\text{liters of solution}}$

Step 3. $M = \dfrac{0.50 \text{ mole}}{6.0 \text{ } L}$

Step 2. moles of solute $= \dfrac{\text{weight in grams}}{\text{weight of 1 mole}}$

Step 4. $M = 0.083 \text{ mole}/L$

weight in grams $= 71$ g
weight of 1 mole of Na $= 23.0 \times 2 = 46.0$ g
weight of 1 mole of S $\;= 32.1 \times 1 = 32.1$ g
weight of 1 mole of O $\;= 16.0 \times 4 = \underline{64.0 \text{ g}}$
142.1 g

moles of solute $= \dfrac{71 \text{ g}}{142.1 \text{ g}}$

moles of solute $= 0.50$ mole
liters of solution $= 6.0$ L

36. Solve the following problem.

0.1 g of $CaCO_3$ is dissolved in 1 L of solution. What is the molarity of this solution?

37. Answers to frame 36.

Step 1. $M = \dfrac{\text{moles of solute}}{\text{liters of solution}}$

Step 2. moles of solute $= \dfrac{\text{weight in grams}}{\text{weight of 1 mole}}$

weight in grams $= 0.1$ g
weight of 1 mole of Ca $= 40.1 \times 1 = 40.1$ g
weight of 1 mole of C $\;= 12.0 \times 1 = 12.0$ g
weight of 1 mole of O $\;= 16.0 \times 3 = \underline{48.0 \text{ g}}$
100.1 g

moles of solute $= \dfrac{0.1 \text{ g}}{100.1 \text{ g}}$

moles of solute $= 0.001$ mole
liters of solution $= 1$ L

Step 3. $M = \dfrac{0.001 \text{ mole}}{1 \text{ } L}$

Step 4. $M = 0.001 \text{ mole}/L$

Part 4. Dilution Formulas

1. Chemicals are frequently purchased in solution form at a standard concentration. If a different concentration is desired, the standard (or stock) solution is diluted. A dilution formula is used that will work with any method of expressing concentration. The dilution formula is as follows.

$$C_1 V_1 = C_2 V_2$$

C_1 = initial concentration of stock solution
V_1 = volume of stock solution that will be diluted
C_2 = concentration of new diluted solution
V_2 = volume of new diluted solution

In the dilution formula the volume of the new solution is symbolized by ——————— .

---

2. The concentration of the stock solution is symbolized in the dilution formula by

——————— .

a. V_1 b. C_2 c. C_1 d. V_2

1. V_2

---

3. The volume of the stock solution is symbolized by ——————— .

a. V_1 b. C_2 c. C_1 d. V_2

2. C_1

---

4. The dilution formula can be arranged to solve for any one of the four quantities listed.

3. V_1

Example: Solve the dilution formula for the new volume (V_2).

Step 1. Write the formula.

$$C_1 V_1 = C_2 V_2$$

Step 2. Divide each side by the unwanted quantity.

$$\frac{C_1 V_1}{C_2} = \frac{C_2 V_2}{C_2}$$

Step 3. Cancel like quantities.

$$\frac{C_1 V_1}{C_2} = \frac{\cancel{C_2} V_2}{\cancel{C_2}} \longrightarrow \frac{C_1 V_1}{C_2} = V_2$$

Step 4. Rewrite the equation.

$$V_2 = \frac{C_1 V_1}{C_2}$$

Solve the dilution formula for C_2.

Step 1. Write the formula.

Step 2. Divide each side by the unwanted quantity.

Step 3. Cancel like quantities.

Step 4. Rewrite the equation.

5. Answers to frame 4.

Step 1. $C_1 V_1 = C_2 V_2$

Step 2. $\dfrac{C_1 V_1}{V_2} = \dfrac{C_2 V_2}{V_2}$

Step 3. $\dfrac{C_1 V_1}{V_2} = \dfrac{C_2 \cancel{V_2}}{\cancel{V_2}} \longrightarrow \dfrac{C_1 V_1}{V_2} = C_2$

Step 4. $C_2 = \dfrac{C_1 V_1}{V_2}$

6. Let us look at a problem involving use of the dilution formula.

Example: What would be the final concentration of 50 mL of a 10%-by-weight solution that was diluted to 100 mL?

Step 1. Write the formula, solving for C_2.

$$C_1 V_1 = C_2 V_2$$

$$\frac{C_1 V_1}{V_2} = \frac{C_2 \cancel{V_2}}{\cancel{V_2}}$$

$$C_2 = \frac{C_1 V_1}{V_2}$$

Step 2. List the data.

$C_1 = 10\%$
$V_1 = 50 \text{ mL}$
$C_2 = ?$
$V_2 = 100 \text{ mL}$

Step 3. Substitute the data into the formula.

$$C_2 = \frac{(10\%)(50\text{mL})}{100 \text{ mL}}$$

Step 4. Do the calculation, first canceling like units.

$$C_2 = \frac{(10\%)(50\cancel{\text{mL}})}{100 \cancel{\text{mL}}}$$

Do the arithmetic.

$$C_2 = 5.0\%$$

Try this problem.

What is the final volume of 2 L of a 1 M solution that has been diluted to a concentration of 0.1 M?

Step 1.

Step 2.

Step 3.

Step 4.

7. Answers to frame 6.

Step 1. $C_1 V_1 = C_2 V_2$

$$V_2 = \frac{C_1 V_1}{C_2}$$

Step 2. $C_1 = 1\,M$
$V_1 = 2\,L$
$C_2 = 0.1\,M$
$V_2 = ?$

Step 3. $V_2 = \frac{(1\,M)(2\,L)}{0.1\,M}$

Step 4. $V_2 = \frac{(1\,\cancel{M})(2\,L)}{0.1\,\cancel{M}}$

$V_2 = 20\,L$

---

8. In practice, you are often required to know how much water must be added to a stock solution to make the diluted solution.

Example: How much water must be added to 20 mL of a 1 M solution to make a 0.1 M solution?

Part A. Solve the dilution formula for V_2.

Step 1. $C_1 V_1 = C_2 V_2$

$$V_2 = \frac{C_1 V_1}{C_2}$$

Step 2. $C_1 = 1\,M$
$V_1 = 20\,mL$
$C_2 = 0.1\,M$
$V_2 = ?$

Step 3. $V_2 = \frac{(1\,M)(20\,mL)}{0.1\,M}$

Step 4. $V_2 = 200\,mL$

Part B. Subtract V_1 from V_2.

200 mL − 20 mL = 180 mL of water to be added

Try this problem.

How much water must be added to 50 mL of a 50%-by-weight solution to make a 10%-by-weight solution?

Part A. Solve the dilution formula for V_2.

Part B. Subtract V_1 from V_2.

---

9. Answers to frame 8.

Part A. Step 1. $V_2 = \frac{C_1 V_1}{C_2}$

Step 2. $C_1 = 50\%$
$V_1 = 50\,mL$
$C_2 = 10\%$
$V_2 = ?$

Step 3. $V_2 = \frac{(50\%)(50\,mL)}{10\%}$

Step 4. $V_2 = 250\,mL$

Part B. $V_2 - V_1 = 250\,mL - 50\,mL = 200\,mL$ of water to be added

10. 20 g of sodium nitrate are dissolved in 80 g of solution. What is the percentage by weight of this solution?

11. Answers to frame 10.

percentage by weight $= \dfrac{\text{grams solute}}{\text{grams solution}} \times 100\%$

grams solute $=$ 20 g
grams solution $=$ 80 g

percentage by weight $= \dfrac{20 \text{ g}}{80 \text{ g}} \times 100\%$

percentage by weight $=$ 25%

12. 10 g of salt are dissolved in 40 mL (40 g) of water. What is the percentage by weight of this solution?

13. Answers to frame 12.

percentage by weight $= \dfrac{\text{grams solute}}{\text{grams solution}} \times 100\%$

grams solute $=$ 10 g
grams solution $=$ 10 g + 40 g = 50 g

percentage by weight $= \dfrac{10 \text{ g}}{50 \text{ g}} \times 100\%$

percentage by weight $=$ 20%

14. 10 mL of alcohol are dissolved in enough water to make 30 mL of solution. What is the percentage by volume of this solution?

15. Answers to frame 14.

$$\text{percentage by volume} = \frac{\text{volume of solute}}{\text{volume of solution}} \times 100\%$$

volume of solute $= 10$ mL
volume of solution $= 30$ mL

$$\text{percentage by volume} = \frac{10 \text{ mL}}{30 \text{ mL}} \times 100\%$$

percentage by volume $= 33\%$

16. 2 moles of salt are dissolved in 5 L of solution. What is the molarity of this solution?

17. Answers to frame 16.

$$M = \frac{\text{moles of solute}}{\text{liters of solution}}$$

moles of solute $= 2$ moles
liters of solution $= 5\ L$

$$M = \frac{2 \text{ moles}}{5\ L}$$

$$M = 0.4 \text{ mole}/L$$

18. 60 g of NaOH are dissolved in 3 L of solution. What is the molarity of this solution?

19. Answers to frame 18.

$$M = \frac{\text{moles of solute}}{\text{liters of solution}}$$

$$\text{moles of solute} = \frac{\text{weight in grams}}{\text{weight of 1 mole}}$$

weight in grams $= 60$ g
weight of 1 mole of Na $= 23.0 \times 1 = 23.0$ g
weight of 1 mole of O $\ = 16.0 \times 1 = 16.0$ g
weight of 1 mole of H $\ = \ \ 1.0 \times 1 = \ \ \underline{1.0}$ g
$$ 40.0 \text{ g}$$

$$\text{moles of solute} = \frac{60 \text{ g}}{40.0 \text{ g}}$$

moles of solute $= 1.5$ moles
liters of solution $= 3\ L$

$$M = \frac{1.5 \text{ moles}}{3\ L}$$

$$M = 0.5 \text{ mole}/L$$

20. Solve the dilution formula for V_1.

21. Answers to frame 20.

$$C_1 V_1 = C_2 V_2$$

$$\frac{\cancel{C_1} V_1}{\cancel{C_1}} = \frac{C_2 V_2}{C_1}$$

$$V_1 = \frac{C_2 V_2}{C_1}$$

22. What would be the final concentration of 20 mL of a 5%-by-weight solution that was diluted to 100 mL?

23. Answers to frame 22.

$$C_1 V_1 = C_2 V_2$$

$$C_2 = \frac{C_1 V_1}{V_2}$$

$C_1 = 5\%$
$V_1 = 20$ mL
$C_2 = ?$
$V_2 = 100$ mL

$$C_2 = \frac{(5\%)(20 \text{ mL})}{100 \text{ mL}}$$

$$C_2 = 1\%$$

24. How much water must be added to 50 mL of a 2 M solution to make a 0.2 M solution?

25. Answers to frame 24.

$$C_1 V_1 = C_2 V_2$$

$$V_2 = \frac{C_1 V_1}{C_2}$$

$C_1 = 2 M$
$V_1 = 50$ mL
$C_2 = 0.2 M$
$V_2 = ?$

$$V_2 = \frac{(2M)(50\text{mL})}{0.2 \, M}$$

$V_2 = 500$ mL
$V_2 - V_1 = 500$ mL $-$ 50 mL $= 450$ mL of water to be added

Part 5. More-Complicated Molarity Problems

1. The units of molarity must always be moles of solute and liters of solution. Frequently, the data in a problem will be given as grams of solute and milliliters of solution. The grams must be converted to moles and the milliliters to liters before the molarity of the solution can be calculated.

   Example: What is the molarity of a solution containing 5.00 g of NaCl in 150 mL of solution?

   Step 1. $M = \dfrac{\text{moles solute}}{\text{liters of solution}}$

   Step 2.   Change the grams of solute (NaCl) to moles and the milliliters of solution to liters.

   a. moles of solute $= \dfrac{\text{wt in g}}{\text{wt/mole}}$

   moles of solute $= \dfrac{5.00 \text{ g}}{58.5 \text{ g/mole}}$ (mol wt of NaCl)

   moles of solute = 0.0855 mole

   b. liters of solution = 150 mL $\times \dfrac{1 \, L}{1000 \text{ mL}} = 0.150 \, L$

   Step 3. $M = \dfrac{0.0855 \text{ mole}}{0.150 \, L}$

   Step 4. $M = 0.570$ mole/L

   This problem can be shortened by using conversion factors.

   $$\underset{(1)}{\dfrac{5.00 \text{ g NaCl}}{150 \text{ mL}}} \times \underset{(2)}{\dfrac{1 \text{ mole NaCl}}{58.5 \text{ g NaCl}}} \times \underset{(3)}{\dfrac{1000 \text{ mL}}{1 \, L}} = 0.570 \text{ mole/}L \text{ or } 0.570 \, M$$

   Conversion factor (1) states the original units. Conversion factor (2) converts grams to moles. Conversion factor (3) converts milliliters to liters.

   Try these problems.

   a. What is the molarity of a solution if 10.0 g of HCl are dissolved in 400 mL of solution?

   b. What is the molarity of a solution if 15.0 g of NaOH are dissolved in 1400 mL of solution?

2. Answers to frame 1.

a. $\dfrac{10.0 \text{ g HCl}}{400 \text{ mL}} \times \dfrac{1 \text{ mole}}{36.5 \text{ g HCl}} \times \dfrac{1000 \text{ mL}}{1 \text{ } L} = 0.685 \text{ mole/}L$

b. $\dfrac{15.0 \text{ g NaOH}}{1400 \text{ mL}} \times \dfrac{1 \text{ mole}}{40.0 \text{ g NaOH}} \times \dfrac{1000 \text{ mL}}{1 \text{ } L} = 0.268 \text{ mole/}L$

---

3. The formula for finding molarity can be rearranged to solve for moles or grams of solute or liters or milliliters of solution.

   Example: How many moles of solute are in 250 mL of a 0.500 M solution?

   Step 1. $M = \dfrac{\text{moles of solute}}{\text{liters solution}}$ or moles of solute $= (M)(\text{liters solution})$

   Step 2. $M = 0.500 \text{ mole/}L$

   $\text{liters solution} = 250 \text{ mL} \times \dfrac{1 \text{ } L}{1000 \text{ mL}} = 0.250 \text{ } L$

   Step 3. moles of solute $= 0.500 \text{ mole/}L \times 0.250 \text{ } L$

   Step 4. moles of solute $= 0.125 \text{ mole}$

   In conversion-factor form, the problem would look like this.

   $$\dfrac{0.500 \text{ mole}}{1 \text{ } L} \times \dfrac{250 \text{ mL}}{1000 \text{ mL/}L} = 0.125 \text{ mole}$$

   Example: How many milliliters of a 0.200 mL/L solution can be prepared from 0.150 mole of a solute?

   Step 1. $M = \dfrac{\text{moles}}{\text{liters}}$ or liters $= \dfrac{\text{moles}}{M}$

   Note: The liters will have to be changed to milliliters in the last step.

   Step 2. $M = 0.200 \text{ mole/}L$
   moles $= 0.150 \text{ mole}$

   Step 3. liters $= \dfrac{0.150 \text{ mole}}{0.200 \text{ mole/}L}$

   Step 4. liters $= 0.750 \text{ } L \times \dfrac{1000 \text{ mL}}{1 \text{ } L} = 750 \text{ mL}$

   In conversion-factor form:

   $$0.150 \text{ mole} \times \dfrac{1 \text{ } L}{0.200 \text{ mole}} \times \dfrac{1000 \text{ mL}}{1 \text{ } L} = 750 \text{ mL}$$

   Try these problems.

   a. How many moles of a solute are in 300 mL of a 0.450 M solution?

   b. How many milliliters of a 0.300 M solution can be prepared from 0.250 moles of solute?

c.   How many moles of KCl are in 100 mL of a 0.110 M solution?

d.   How many milliliters of a 0.180 M solution can be prepared from 0.100 moles of $CaCl_2$?

---

4. Answers to frame 3.

a. $\dfrac{0.450 \text{ mole}}{1 \, L} \times \dfrac{300 \text{ mL}}{1000 \text{ mL/}L} = 0.135 \text{ mole}$

b. $0.250 \text{ mole} \times \dfrac{1 \, \ell}{0.300 \text{ mole}} \times \dfrac{1000 \text{ mL}}{1 \, L} = 833 \text{ mL}$

c. $\dfrac{0.110 \text{ mole}}{1 \, L} \times \dfrac{100 \text{ mL}}{1000 \text{ mL/}L} = 0.0110 \text{ moles}$

d. $0.100 \text{ mole} \times \dfrac{1 \, L}{0.180 \text{ mole}} \times \dfrac{1000 \text{ mL}}{1 \, L} = 556 \text{ mL}$

---

5. The same type of problem illustrated in frame 2 can be solved for grams of solute instead of moles.

Example: How many grams of NaOH are needed to prepare 400 mL of a 0.215 M solution?

Step 1. $M = \dfrac{\text{moles (NaOH)}}{\text{liters solution}}$ or   moles NaOH = (M)(liters)

In the last step, the moles of NaOH will be converted to grams.

Step 2. M = 0.215 moles/L

liters = $400 \text{ mL} \times \dfrac{1 \, L}{1000 \text{ mL}} = 0.400 \, L$

Step 3.   moles NaOH = 0.215 moles/L $\times$ 0.400 L

Step 4.   moles NaOH = 0.086 mole

g of NaOH = $0.086 \text{ mole} \times \dfrac{40.0 \text{ g}}{\text{mole}} = 3.44 \text{ g}$

In conversion-factor form:

$$\dfrac{0.215 \text{ moles}}{L} \times \dfrac{400 \text{ mL}}{1000 \text{ mL/}L} \times \dfrac{40.0 \text{ g}}{1 \text{ mole}} = 3.44 \text{ g}$$

Example: How many milliliters of a 0.450 M solution can be prepared from 20.5 g of KOH?

Step 1. $M = \dfrac{\text{moles KOH}}{\text{liters}}$ or   liters = $\dfrac{\text{moles KOH}}{M}$

Change to mL in the last step.

Step 2. M = 0.450 moles/liter

moles KOH = $\dfrac{20.5 \text{ g}}{56.1 \text{ g/mole}} = 0.365 \text{ mole}$

Step 3.   liters $= \dfrac{0.365 \text{ moles}}{0.450 \text{ moles/liter}}$

Step 4.   liters $= 0.811$

milliliters $= 0.811 \times \dfrac{100 \text{ mL}}{1 \text{ } L} = 811 \text{ mL}$

In conversion-factor form:

$$20.5 \text{ g KOH} \times \frac{1 \text{ mole KOH}}{56.1 \text{ g KOH}} \times \frac{1 \text{ } L}{0.450 \text{ mole KOH}} \times \frac{1000 \text{ mL}}{1 \text{ } L} = 812 \text{ mL}$$

There is a slight difference in answers due to rounding off.

Try these problems.

a.   How many grams of $MgCl_2$ are needed to prepare 200 mL of a 0.450 M solution?

b.   How many milliliters of a 0.850 M solution can be prepared from 16.5 g of NaCl?

c.   How many grams of $Ca(OH)_2$ are needed to prepare 300 mL of a 0.115 M solution?

d.   How many milliliters of a 0.650 M solution can be prepared from 5.12 g of $Mg(OH)_2$?

---

6. Answers to frame 5.

a. $\dfrac{0.450 \text{ mole}}{1 \text{ } L} \times \dfrac{200 \text{ mL}}{1000 \text{ mL/}L} \times \dfrac{95.3 \text{ g}}{1 \text{ mole}} = 8.58 \text{ g}$

b. $16.5 \text{ g} \times \dfrac{1 \text{ mole NaCl}}{58.5 \text{ g NaCl}} \times \dfrac{1 \text{ } L}{0.850 \text{ mole}} \times \dfrac{1000 \text{ mL}}{1 \text{ } L} = 332 \text{ mL}$

c. $\dfrac{0.115 \text{ mole}}{1 \text{ } L} \times \dfrac{300 \text{ mL}}{1000 \text{ mL/}L} \times \dfrac{74.1 \text{ g}}{1 \text{ mole}} = 2.56 \text{ g}$

d. $5.12 \text{ g} \times \dfrac{1 \text{ mole Mg(OH)}_2}{58.3 \text{ g Mg(OH)}_2} \times \dfrac{1 \text{ } L}{0.650 \text{ mole}} \times \dfrac{1000 \text{ mL}}{1 \text{ } L} = 135 \text{ mL}$

Part 6. Normality and Equivalent Weight

1. The concentration of a solution can be expressed in units of normality (N). The formula for the normality of a solution is:

$$\text{Normality } (N) = \frac{\text{number of equivalent weights of solute}}{1 \text{ liter of solution}}$$

The equivalent weight or equivalent of a solute is related to the weight of one mole of the solute in the following manner.

$$\text{Equivalent wt of a solute } = \frac{\text{wt of one mole of solute}}{n}$$
$$n = 1, 2, 3, \ldots$$

A series of neutralization reactions illustrates this concept.

A. $\text{HCl} + \text{NaOH} \longrightarrow \text{NaCl} + \text{H}_2\text{O}$
　　　1 mole　　1 mole
　　(36.5 g/mole)　(40.0 g/mole)

The acid and base react in a one-mole-to-one-mole ratio, or 36.5 g of HCl react with 40.0 g of NaOH. These weights are equivalent weights, as well as the weight of one mole each, because they react in a one-to-one ratio.

B. $\text{H}_2\text{SO}_4 + 2\text{NaOH} \longrightarrow \text{Na}_2\text{SO}_4 + 2\text{H}_2\text{O}$
　　　1 mole　　2 mole
　(98.1 g/mole)　(40.0 g/mole)

The acid and base react in a one-mole-to-two-mole ratio. That weight of acid (H_2SO_4) that will react with one mole (40.0 g) of NaOH is:

2 moles NaOH react with 1 mole H_2SO_4 or

1 mole NaOH reacts with $\dfrac{1 \text{ mole H}_2\text{SO}_4}{2}$

40.0 g/equiv wt of NaOH $\dfrac{98.1}{2} = 49.0$ g/equiv wt of H_2SO_4

C. $\text{H}_3\text{PO}_4 + 3\text{NaOH} \longrightarrow \text{Na}_3\text{PO}_4 + 3\text{H}_2\text{O}$
　　　1 mole　　3 mole
　(98.1 g/mole)　(40.0 g/mole)

3 moles NaOH react with 1 mole H_3PO_4 or

1 mole NaOH reacts with $\dfrac{1 \text{ mole H}_3\text{PO}_4}{3}$

40.0 g/equiv wt of NaOH $\dfrac{98.0 \text{ g}}{3} = 32.7$ g/equiv wt of H_3PO_4

Notice that, in the reactions,

A. 1 mole of HCl gives up 1 mole of H^+ ions

　　$\text{HCl} \longrightarrow \text{H}^+ + \text{Cl}^-$

B. 1 mole of H_2SO_4 gives up 2 moles of H^+ ions

　　$\text{H}_2\text{SO}_4 \longrightarrow 2\text{H}^+ + \text{SO}_4^{2-}$

C. 1 mole of H_3PO_4 gives up 3 moles of H^+ ions

　　$\text{H}_3\text{PO}_4 \longrightarrow 3\text{H}^+ + \text{PO}_4^{3-}$

If we divide the weight of one mole of the acid by the number of moles of H^+ ions produced, we get the equivalent weight.

A. Equivalent wt of HCl $= \dfrac{36.5 \text{ g/mole}}{1 \text{ mole}} = 36.5$ g

B. Equivalent wt of $H_2SO_4 = \dfrac{98.1 \text{ g/mole}}{2 \text{ moles}} = 49.0$ g

C. Equivalent wt of $H_3PO_4 = \dfrac{98.0 \text{ g/mole}}{3 \text{ moles}} = 32.7$ g

Acids were used in the illustrations, but the equivalent weight of any substance can be defined as that weight which will react with or contain 1 mole of H^+ ions.

Calculate the equivalent weight of H_2CO_3 in the following reaction.

$$H_2CO_3 + 2NaOH \longrightarrow Na_2CO_3 + 2H_2O$$

---

2. Answers to frame 1.

Wt of one mole $H_2CO_3 = 62.0$ g
Number of moles of H^+ produced $= 2$

equivalent wt of $H_2CO_3 = \dfrac{62.0 \text{ g}}{2} = 31.0$ g

---

3. The equivalent weight of a base can be calculated by finding what weight of the base will react with 1 mole of HCl.

Example: $2HCl + Ca(OH)_2 \longrightarrow CaCl_2 + 2H_2O$
    2 moles   1 mole
    (36.5 g)   (74.1 g)

2 moles of HCl react with 1 mole $Ca(OH)_2$ or

1 mole HCl reacts with $\dfrac{1 \text{ mole Ca(OH)}_2}{2}$

36.5 g/equiv wt of HCl $\dfrac{74.1 \text{ g}}{2} = 37.1$ g/equiv wt of $Ca(OH)_2$

What is the equivalent wt of $Al(OH)_3$ in the following reaction?

$$3HCl + Al(OH)_3 \longrightarrow AlCl_3 + 3H_2O$$

4. The number of equivalent weights can be found using the following formula.

$$\text{Number of equiv wt} = \frac{\text{wt in g}}{\text{wt/equiv wt}}$$

$3. \dfrac{78.0 \text{ g}}{3} = 26.0 \text{ g}$

Example: The equivalent weight of H_2SO_4 is 49.0 g. How many equivalents are in 150 g of H_2SO_4?

Step 1. Number of equivalents $= \dfrac{\text{wt in g}}{\text{wt/equiv}}$

Step 2. wt in g $= 150$ g
wt/equiv $= 49.0$ g/equiv

Step 3. Number of equivalents $= \dfrac{150 \text{ g}}{49.0 \text{ g/equiv}}$

Step 4. Number of equivalents $= 3.06$ equiv   or

$$150 \text{ g } H_2SO_4 \times \frac{1 \text{ equiv}}{49.0 \text{ g } H_2SO_4} = 3.06 \text{ equiv}$$

Try these problems.

a. The equivalent weight of $Ca(OH)_2$ is 37.1 g. How many equivalents are there in 10.5 g of $Ca(OH)_2$?

b. The equivalent weight of H_3PO_4 is 32.7 g. How many equivalents are in 88.5 g of H_3PO_4?

5. Answers to frame 4.

a. $10.5 \text{ g } Ca(OH)_2 \times \dfrac{1 \text{ equiv}}{37.1 \text{ g } Ca(OH)_2} = 0.283$ equiv

b. $88.5 \text{ g of } H_3PO_4 \times \dfrac{1 \text{ equiv}}{32.7 \text{ g } H_3PO_4} = 2.71$ equiv

6. The normality (N) of a solution can be found using the following two formulas.

$$\text{Number of equivalents} = \frac{\text{wt in g}}{\text{wt/equiv}}$$

$$\text{Normality } (N) = \frac{\text{number of equivalents}}{\text{liter of solution}}$$

Example: What is the normality of an H_2SO_4 solution containing 30.4 g of H_2SO_4 in 350 mL of solution?

Part A. Find the number of equivalents of H_2SO_4.

1. Number of equivalents $= \dfrac{\text{wt in g}}{\text{wt/equiv}}$

2. wt in g = 30.4 g

$$\text{wt/equiv} = \frac{98.1 \text{ g}}{2 \text{ equiv}} = 49.1 \text{ g/equiv}$$

3. Number of equiv = $\dfrac{30.4 \text{ g}}{49.1 \text{ g/equiv}}$

4. Number of equiv = 0.619 equiv

Part B.   Use the answer from Part A to solve for the normality.

1. $N = \dfrac{\text{number of equiv}}{\text{liters of solution}}$

2. number of equiv = 0.619 equiv

liters of solution = 350 mL $\times \dfrac{1 \text{ L}}{1000 \text{ mL}} = 0.350 \text{ L}$

3. $N = \dfrac{0.619 \text{ equiv}}{0.350 \text{ L}}$

4. $N = 1.77 \text{ equiv/L}$

In conversion-factor form:

$$N = \frac{30.4 \text{ g}}{0.350 \text{ L}} \times \frac{1 \text{ equiv}}{49.1 \text{ g}} = 1.77 \text{ equiv/L}$$

Try these problems.

a. What is the normality of a $Ca(OH)_2$ solution containing 0.860 g of $Ca(OH)_2$ in 255 mL of solution?

b. What is the normality of a H_3PO_4 solution containing 5.75 g of H_3PO_4 in 150 mL of solution?

---

7. Answers to frame 6.

a. Number of equivalents = $\dfrac{0.860 \text{ g}}{37.1 \text{ g/equiv}} = 0.0232 \text{ equiv}$

$N = \dfrac{0.0232 \text{ equiv}}{0.225 \text{ L}} = 0.103 \text{ equiv/L}$

b. Number of equivalents = $\dfrac{5.75 \text{ g}}{32.7 \text{ g/equiv}} = 0.176 \text{ equiv}$

$N = \dfrac{0.176 \text{ equiv}}{0.150 \text{ L}} = 1.17 \text{ equiv/L}$

Part 7. Titration

1. During a titration, a solution of known concentration is added to a solution of unknown concentration until the reaction between the solutes is complete. At the completion point of the reaction, three quantities are known, and a fourth can be calculated.

Known

C_1 = concentration of known solution (M or N)
V_1 = volume of known solution (liters or mL)
V_2 = volume of unknown solution (liters or mL)

Unknown

C_2 = concentration of unknown solution

$$C_2 = \frac{C_1 V_1}{V_2}$$

Example: 15.1 mL of 0.350 M NaOH solution is required to titrate 25.0 mL of HCl solution. What is the molarity of the HCl solution? The reaction is:

$$HCl + NaOH \longrightarrow NaCl + H_2O$$

Step 1. $C_2 = \dfrac{C_1 V_1}{V_2}$

Step 2. C_1 = 0.350 M NaOH
 V_1 = 15.1 mL NaOH
 V_2 = 25.0 mL HCl

Step 3. $C_2 = \dfrac{(0.350\ M)(15.1\ mL)}{25.0\ mL}$

Step 4. C_2 = 0.211 M

Molarity was used in this problem because HCl and NaOH react in a 1:1 mole ratio. Normality is used when the mole ratio in not 1:1.

Example: 30.4 mL of 0.250 M H_2SO_4 solution is required to titrate 20.0 mL of NaOH solution. What is the molarity of the NaOH solution? The reaction is:

$$H_2SO_4 + 2NaOH \longrightarrow Na_2SO_4 + 2H_2O$$

Step 1. $C_2 = \dfrac{C_1 V_1}{V_2}$

Step 2. C_1 = 0.250 M H_2SO_4
 V_1 = 30.4 mL H_2SO_4
 V_2 = 20.0 mL NaOH

Step 3. $C_2 = \dfrac{(0.250\ M)(30.4\ mL)}{20.0\ mL}$

Step 4. C_2 = 0.380 M

Try these problems.

a. 20.6 mL of 0.150 M NaOH solution is needed to titrate 16.8 mL of HCl solution. What is the molarity of the HC1 solution?

b. 14.6 mL of 0.150 N H_2SO_4 solution is needed to titrate 24.8 mL of NaOH solution. What is the normality of the NaOH solution?

2. Answers to frame 1.

a. $C_2 = \dfrac{(0.150\ M)(20.6\ \text{mL})}{16.8\ \text{mL}} = 0.184\ M$

b. $C_2 = \dfrac{(0.150\ N)(14.6\ \text{mL})}{24.8\ \text{mL}} = 0.0883\ N$

Part 8. The pH of a Solution

1. Acids produce H^+ ions when in solution. A base produces OH^- ions, which neutralize the H^+ ions and decrease their number. The strength of an acid or base is determined by measuring the amount of hydrogen ions present. The common unit used to measure the strength of an acid or base is molarity (M).

The molarity (M) of an acid or base is stated in power-of-ten form. The power-of-ten form is difficult to work with, so the strength of an acid or base is expressed as the logarithm of the power-of-ten form. The log form is called the pH of the acid or base.

$$pH = -\log_{10}[H^+]$$

pH = the negative log of the hydrogen ion concentration. The minus sign changes the answer to a positive number.

$[H^+]$ = the concentration of the hydrogen ion in molarity (M).

Example: What is the pH of a solution that has a concentration of $5.5 \times 10^{-4}\ M$ of hydrogen ions?

Step 1. Substitute into the pH formula.

$$pH = -\log(5.5 \times 10^{-4})$$

Step 2. Logs are exponents and, when multiplied, exponents are added. The formula

$$pH = -\log(5.5 \times 10^{-4})$$

can be rewritten as:

$$pH = -(\log 5.5 + \log 10^{-4})$$

Use your calculator to find the log of 5.5.

$\log 5.5 = 0.74$

$\log 10^{-4} = -4.00$

$pH = -(0.74 + -4.00)$

$pH = -(-3.26) = 3.26$

Find the pH of these solutions, given their concentration in molarity.

a. $1.6 \times 10^{-2} M$ _____   b. $2.7 \times 10^{-4} M$ _____

---

2. Answers to frame 1.

a. 1.80   b. 3.57

---

3. The pH scale goes from 0 to 14. A solution with a pH of 7 is neutral. A solution with a pH between 0 and 6.99 is acidic. A solution with a pH between 7.01 and 14 is basic.

pH Scale

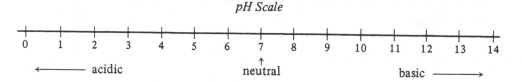

The above scale can be used to solve the following problem.
A solution has a hydrogen ion concentration of 3.75×10^{-8} M. Is the solution acidic or basic?

Step 1.   Substitute into the pH formula.

$$pH = -\log (3.75 \times 10^{-8})$$

Step 2.   Solve for the pH.

$$pH = -\log (3.75 \times 10^{-8})$$
$$pH = -(\log 3.75 + \log 10^{-8})$$
$$\log 3.75 = 0.57 \quad \text{and} \quad \log 10^{-8} = -8.00$$
$$pH = -(0.57 + -8.00)$$
$$pH = -(-7.43) = 7.43$$

From the pH scale we can see that a solution with a pH of 7.43 would be basic.

Find the pH of the following solutions, given the hydrogen ion concentrations, and state whether the solution is acidic or basic.

a. $2.71 \times 10^{-5} M$ _____   b. $8.79 \times 10^{-11} M$ _____

c. $6.04 \times 10^{-7} M$ _____   d. $6.87 \times 10^{-6} M$ _____

---

4. Answers to frame 3.

a.   4.57–acidic   b.   10.1–basic
c.   6.22–acidic   d.   5.16–acidic

---

5. What is the molarity of a solution containing 10.0 g of $NaNO_3$ in 250 mL of solution?

6. Answers to frame 5.

$$\frac{10.0 \text{ g NaNO}_3}{250 \text{ mL}} \times \frac{1 \text{ mole NaNO}_3}{85.0 \text{ g NaNO}_3} \times \frac{1000 \text{ mL}}{1 L} = 0.471 \text{ mole/}L$$

7. What is the molarity of a solution containing 0.460 g of $CaCl_2$ in 50.5 mL of solution?

8. Answers to frame 7.

$$\frac{0.46 \text{ g CaCl}_2}{50.5 \text{ mL}} \times \frac{1 \text{ mole CaCl}_2}{111.1 \text{ g CaCl}_2} \times \frac{1000 \text{ mL}}{1 L} = 0.820 \text{ mole/}L$$

9. How many moles of solute are in 150 mL of a 0.115 M solution?

10. Answers to frame 9.

$$\frac{0.115 \text{ mole}}{1 L} \times \frac{150 \text{ mL}}{1000 \text{ mL/}L} = 0.0173 \text{ mole}$$

11. How many milliliters of a 1.50 M solution can prepared from 1.10 moles of NaOH?

12. Answers to frame 11.

$$1.10 \text{ mole} \times \frac{1 L}{1.50 \text{ mole}} \times \frac{1000 \text{ mL}}{1 L} = 733 \text{ mL}$$

13. How many grams of $AgNO_3$ are needed to prepare 100 mL of a 0.150 M solution?

14. Answers to frame 13.

$$\frac{0.150 \text{ mole}}{1 L} \times \frac{100 \text{ mL}}{1000 \text{ mL/}L} \times \frac{169.9 \text{ g AgNO}_3}{1 \text{ mole}} = 2.55 \text{ g}$$

15. How many milliliters of a 0.170 M solution can be prepared from 10.3 g of KCl?

16. Answers to frame 15.

$$10.3 \text{ g} \times \frac{1 \text{ mole KCl}}{74.6 \text{ g KCl}} \times \frac{1 \, L}{0.170 \text{ mole}} \times \frac{1000 \text{ mL}}{1 \, L} = 812 \text{ mL}$$

17. Calculate the equivalent weight of the acid in the following reactions.

   a. $H_2SO_3 + 2NaOH \longrightarrow Na_2SO_3 + 2H_2O$

   b. $H_2S + 2NaOH \longrightarrow Na_2S + 2H_2O$

18. Answers to frame 17.

   a. $\dfrac{82.1 \text{ g } H_2SO_3}{2} = 41.1 \text{ g}$ b. $\dfrac{34.1 \text{ g } H_2S}{2} = 17.1 \text{ g}$

19. Calculate the equivalent weight of the base in the following reactions.

   a. $2HCl + Mg(OH)_2 \longrightarrow CaCl_2 + 2H_2O$

   b. $2HCl + Fe(OH)_2 \longrightarrow FeCl_2 + 2H_2O$

20. Answers to frame 19.

   a. $\dfrac{58.3 \text{ g } Mg(OH)_2}{2} = 29.2 \text{ g}$ b. $\dfrac{89.8 \text{ g } Fe(OH)_2}{2} = 44.9 \text{ g}$

21. The equivalent weight of $Mg(OH)_2$ is 29.2 g. How many equivalents are in 26.4 g of $Mg(OH)_2$?

22. Answers to frame 21.

   $$26.4 \text{ g} \times \frac{1 \text{ equiv}}{29.2 \text{ g}} = 0.904 \text{ equiv}$$

23. What is the normality of an H_2SO_4 solution containing 8.25 g of H_2SO_4 in 500 mL of solution?

---

24. Answers to frame 23.

$$\text{Number of equiv} = 8.25 \text{ g} \times \frac{1 \text{ equiv}}{49.1 \text{ g}} = 0.168$$

$$N = \frac{0.168 \text{ equiv}}{0.550 \text{ L}} = 0.305 \text{ equiv/L}$$

---

25. What is the normality of an $Mg(OH)_2$ solution containing 3.50 g of $Mg(OH)_2$ in 200 mL of solution?

---

26. Answers to frame 25.

$$\text{Number of equiv} = 3.50 \text{ g} \times \frac{1 \text{ equiv}}{29.2 \text{ g}} = 0.120$$

$$N = \frac{0.120 \text{ equiv}}{0.200 \text{ L}} = 0.600 \text{ equiv/L}$$

---

27. 18.7 mL of 0.250 M NaOH solution is needed to titrate 22.4 mL of HCl solution. What is the molarity of the HCl solution?

---

28. Answers to frame 27.

$$C_2 = \frac{(0.250 \ M)(18.7 \text{ mL})}{(22.4 \text{ mL})} = 0.209 \ M$$

---

29. 30.2 mL of 0.320 N H_2SO_4 solution is needed to titrate 28.6 mL of NaOH solution. What is the normality of the NaOH solution?

---

30. Answers to frame 29.

$$C_2 = \frac{(0.320 \ N)(30.2 \text{ mL})}{(28.6 \text{ mL})} = 0.338 \ N$$

---

31. Find the pH of the following solutions, given the hydrogen ion concentration, and state whether the solution is acidic or basic.

   a. $3.75 \times 10^{-4} M$ _____

   b. $8.25 \times 10^{-10} M$ _____

32. Answers to frame 31.

   a.   3.43—acidic   b.   9.08—basic

THE CONCENTRATION OF SOLUTIONS
EVALUATION TEST 1
PART 1–PART 3 (FRAME 21)

1. Solutions are _____ .

    a. mixtures   b. pure substances

2. That which is dissolved is the _____ .

    a. solute   b. solvent   c. solution

3. That which does the dissolving is the _____ .

    a. solute   b. solvent   c. solution

4. When sugar is dissolved in water, the sugar is the _____ .

    a. solute   b. solvent   c. solution

5. When alcohol is dissolved in water, the water is the _____ .

    a. solute   b. solvent   c. solution

6. A solution that contains a small amount of solute in a large amount of solution is called _____ .

    a. concentrated   b. dilute
    c. saturated   d. unsaturated

7. A solution that contains all of the solute it can at a given temperature is called _____ .

    a. concentrated   b. dilute
    c. saturated   d. unsaturated

8. A solution that can dissolve more solute is called _____ .

    a. concentrated   b. dilute
    c. saturated   d. unsaturated

9. The formula for percentage by weight is _____ .

10. In a percentage-by-weight problem, which two of the following must you know in order to solve the problem?

    a. moles of solute   b. grams of solute
    c. grams of solution   d. volume of solution

11. Given the number of grams of solute and the number of grams of solvent, the formula for finding the grams of solution is as follows.

    grams of solution = _____

12. If 30 g of salt are dissolved in 50 g of water,

    a. grams solute = _____ ,

    b. grams solvent = _____ ,

    c. grams solution = _____ .

Answer column:

1. _____
2. _____
3. _____
4. _____
5. _____
6. _____
7. _____
8. _____
9. _____
10. _____
11. _____
12. a. _____
    b. _____
    c. _____

THE CONCENTRATION OF SOLUTIONS
EVALUATION TEST 2
PART 3 (FRAME 22)–PART 3 (FRAME 37)

1. Which of the following is the formula needed to solve for percentage by volume?

    a. $\dfrac{\text{moles solute}}{\text{liters solution}}$ b. $\dfrac{\text{volume solute}}{\text{volume solution}} \times 100\%$ c. $\dfrac{\text{g solute}}{\text{g solution}} \times 100\%$

2. Which of the following is the formula needed to solve for molarity?

    a. $\dfrac{\text{moles solute}}{\text{liters solution}}$ b. $\dfrac{\text{volume solute}}{\text{volume solution}} \times 100\%$ c. $\dfrac{\text{g solute}}{\text{g solution}} \times 100\%$

3. 5.0 g of NaCl are dissolved in 40 g of solution. What is the percentage by weight of this solution?

4. 20 g of NaOH are dissolved in 80 mL (80 g) of water. What is the percentage by weight of this solution?

5. 20 mL of alcohol are dissolved in enough water to make 200 mL of solution. What is the percentage by volume of this solution?

6. 4 moles of salt are dissolved in 5 L of solution. What is the molarity of this solution?

1. _____

2. _____

3. _____

4. _____

5. _____

6. _____

THE CONCENTRATION OF SOLUTIONS
EVALUATION TEST 3
PART 4

1. Which of the following is the formula needed to convert grams to moles?

    a. $\dfrac{\text{grams solute}}{\text{grams solution}} \times 100\%$

    b. grams solute + grams solvent

    c. $\dfrac{\text{weight in grams}}{\text{weight of 1 mole}}$

2. In the dilution formula, C_1 represents _____ .

    a. concentration of diluted solution
    b. concentration of stock solution
    c. water added

3. 0.5 g of $CaCO_3$ is dissolved in 2 L of solution. What is the molarity of this solution?

4. Solve the dilution formula for C_2.

5. What will be the final concentration of 80 mL of a 0.8 M solution that is diluted to 320 mL?

6. How much water must be added to 20 mL of a 20%-by-weight solution to make a 5%-by-weight solution?

1. _____

2. _____

3. _____

4. _____

5. _____

6. _____

THE CONCENTRATION OF SOLUTIONS
EVALUATION TEST 4
PART 5

1.   What is the molarity of a solution containing 15.0 g of KBr in 175 mL of solution?

2.   How many moles of solute are in 600 mL of a 0.280 M solution?

3.   How many milliliters of a 0.850 M solution can be prepared from 0.625 mole of KOH?

4.   How many grams of $NaNO_3$ are needed to prepare 50.0 mL of a 0.550 M solution?

5.   How many milliliters of a 0.320 M solution can be prepared from 10.5 g of NH_4Cl?

THE CONCENTRATION OF SOLUTIONS
EVALUATION TEST 5
PART 6–PART 8

1. Calculate the equivalent weight of the acid in the following reaction.

$$H_2SO_4 + 2Ca(OH)_2 \longrightarrow CaSO_4 + 2H_2O$$

2. The equivalent weight of HNO_3 is 63.0 g. How many equivalents are in 21.0 g of HNO_3?

3. What is the normality of an H_3PO_4 solution containing 5.60 g of H_3PO_4 in 100 mL of solution?

4. 21.9 mL of a 0.475 M HCl solution is needed to titrate 28.1 mL of a basic solution. What is the molarity of the basic solution?

5. Find the pH of the following solutions, given the hydrogen ion concentration, and state whether the solution is acidic or basic.

   a. 6.75×10^{-7} M   b. 8.95×10^{-12} M

UNIT 14
The Gas Laws

In this unit you will learn to measure the effects of temperature and pressure on the volume of a gas. You will learn the units used in measuring pressure, the method for calculating the effect of pressure on the volume of a gas using Boyle's Law, the Kelvin system of temperature measurement, the method for calculating the effect of temperature on the volume of a gas using Charles' Law, and the method for calculating the combined effect of pressure and temperature on the volume of a gas using the general gas law. You will use the ideal-gas equation to solve gas-law problems. You will solve stoichiometry problems involving gases.

Part 1. Measuring the Pressure of Gases

1. There are many units used to measure pressure. We shall look at only two of these: the atmosphere and the torr. The simplest unit of pressure measurement is the atmosphere, abbreviated as atm. 1 atm is the average pressure of the atmosphere at sea level. Pressure twice as great as the average pressure at sea level would equal 2 atm.

   Pressure one-half as great as the average pressure at sea level would equal _____ atm.

2. In a barometer 1 atm will push mercury 760 mm high into the vacuum tube. We say that the pressure is "760 torr" when the column of mercury is this high. The torr, the second unit used to measure pressure, was named in honor of Torricelli, who invented the barometer.

   1. 0.5 atm

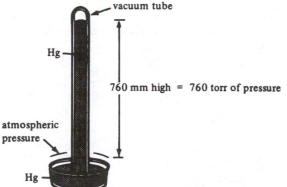

One-half the pressure of the atmosphere (0.5 atm) would push the mercury only 380 mm high (1/2 × 760 mm) into the vacuum tube. This pressure would correspond

to _____ torr.

3. The abbreviation for the unit of measure called *atmosphere* is _____ .

4. In measuring the height of a column of mercury in a vacuum tube, the term used instead of *millimeters of mercury* is the _____ .

5. One-fourth atm is equivalent to _____ torr.

6. Three times the average atmospheric pressure is equal to _____ atm.

2. 380

3. atm

4. torr

5. 190

Part 2. The Effect of Pressure on Gases: Boyle's Law

1. The volume of a gas is inversely proportional to the pressure at constant temperature. This means that, when pressure increases, the volume decreases, and vice versa. Thus, when the pressure on a quantity of gas is doubled, the volume decreases to one-half. If the pressure on a quantity of gas is reduced to one-half, the volume will double. Robert Boyle was the man who first discovered this relationship, and the relationship has come to be known as *Boyle's Law*.

   If the pressure on a quantity of gas is tripled, the volume will decrease to _____

   _____ .

6. 3

2. Boyle's Law is best expressed with the following equation.

$$P_1 V_1 = P_2 V_2$$

P_1 = initial pressure on the gas (in atm or torr)
V_1 = initial volume of the gas (in L or mL)
P_2 = new pressure on the gas (in atm or torr)
V_2 = new volume of the gas (in L or mL)

In the equation for Boyle's Law, the new volume is labeled _____ .

1. one-third

3. In the equation for Boyle's Law, the initial pressure on the gas is labeled _____ .

   a. P_2 b. V_1 c. V_2 d. P_1

2. V_2

4. The new pressure on the gas, in the equation for Boyle's Law, is labeled _____ .

   a. P_2 b. V_1 c. V_2 d. P_1

3. d

5. The Boyle's Law equation can be arranged to solve for any one of the four quantities listed. Generally, we solve for the new volume.

   Example: Solve for V_2.

   1. Write the formula.

   $$P_1 V_1 = P_2 V_2$$

   2. Divide each side by the unwanted quantity.

   $$\frac{P_1 V_1}{P_2} = \frac{P_2 V_2}{P_2}$$

4. a

3. Divide like quantities.

$$\frac{P_1 V_1}{P_2} = \frac{P_2 V_2}{P_2}$$

4. Rewrite the equation.

$$V_2 = \frac{P_1 V_1}{P_2}$$

Solve the Boyle's Law equation for P_2.

1.

2.

3.

4.

---

6. Answers to frame 5.

1. $P_1 V_1 = P_2 V_2$

2. $\dfrac{P_1 V_1}{V_2} = \dfrac{P_2 V_2}{V_2}$

3. $\dfrac{P_1 V_1}{V_2} = \dfrac{P_2 V_2}{V_2}$

4. $P_2 = \dfrac{P_1 V_1}{V_2}$

---

7. Let's solve a Boyle's Law problem.

Example: 300 mL of a gas are under a pressure of 4.00 atm. What would be the volume of the gas at a pressure of 3.00 atm?

1. Write the formula.

$$P_1 V_1 = P_2 V_2 \longrightarrow V_2 = \frac{P_1 V_1}{P_2}$$

2. List the data.

$P_1 = 4.00 \text{ atm}$
$V_1 = 300 \text{ mL}$
$P_2 = 3.00 \text{ atm}$
$V_2 = ?$

3. Substitute the data into the formula.

$$V_2 = \frac{(4.00 \text{ atm})(300 \text{ mL})}{3.00 \text{ atm}}$$

4. Do the calculation.

a. Cancel like units.

$$V_2 = \frac{(4.00 \ \cancel{\text{atm}})(300 \text{ mL})}{3.00 \ \cancel{\text{atm}}}$$

b. Do the arithmetic.

$$V_2 = 400 \text{ mL}$$

Solve the following Boyle's Law problem.

8.0 L of a gas are under 12.4 atm of pressure. What would the volume of the gas be at 6.2 atm of pressure?

1.

2.

3.

4.

---

8. Answers to frame 7.

1. $P_1 V_1 = P_2 V_2 \longrightarrow V_2 = \dfrac{P_1 V_1}{P_2}$

2. $P_1 = 12.4$ atm
   $V_1 = 8.0\ L$
   $P_2 = 6.2$ atm
   $V_2 = ?$

3. $V_2 = \dfrac{(12.4\ \text{atm})(8.0\ L)}{6.2\ \text{atm}}$

4. $V_2 = \dfrac{(12.4\ \cancel{\text{atm}})(8.0\ L)}{6.2\ \cancel{\text{atm}}}$

   $V_2 = 16\ L$

---

9. Try this problem.

200 mL of gas are under a pressure of 600 torr. What would the volume of gas be at a pressure of 800 torr?

1.

2.

3.

4.

---

10. Answers to frame 9.

1. $P_1 V_1 = P_2 V_2 \longrightarrow V_2 = \dfrac{P_1 V_1}{P_2}$

2. $P_1 = 600$ torr
   $V_1 = 200$ mL
   $P_2 = 800$ torr
   $V_2 = ?$

3. $V_2 = \dfrac{(600\ \text{torr})(200\ \text{mL})}{800\ \text{torr}}$

4. $V_2 = \dfrac{(600\ \cancel{\text{torr}})(200\ \text{mL})}{800\ \cancel{\text{torr}}}$

   $V_2 = 150$ mL

Part 3. Measuring the Temperature of Gases

1. The temperature of gases is measured in the Kelvin scale of measurement. The Kelvin scale is compared to the Celsius scale below.

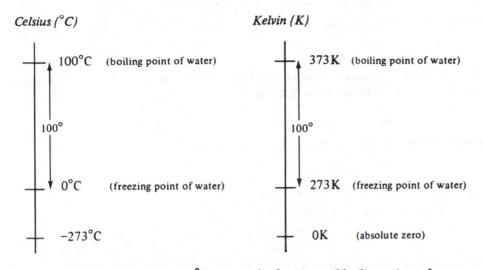

Celsius (°C)   *Kelvin (K)*

100°C (boiling point of water)   373 K (boiling point of water)

100°   100°

0°C (freezing point of water)   273 K (freezing point of water)

−273°C   0 K   (absolute zero)

Note that both systems have 100° between the freezing and boiling points of water. The zero point on the Kelvin scale is the point at which there is no heat left in a system, hence the name *absolute zero*.

The abbreviation for the Kelvin system of temperature measurement is _____.

2. The freezing point of water in the Kelvin system is _____ K.

3. 0 K is also known as _____ .

4. Celsius temperatures are easily converted to Kelvin temperatures with the use of the following formula.

$$K = °C + 273$$

Example: 20°C is what temperature on the Kelvin scale?

1. K = °C + 273
2. °C = 20
3. K = 20 + 273
4. K = 293

Try this problem.

−10°C is what temperature on the Kelvin scale?

1.

2.

3.

4.

---

1. K

2. 273

3. absolute zero

5. Answers to frame 4.

   1.   K = °C + 273
   2.   °C = −10

   3.   K = (−10) + 273
   4.   K = 263

6. Change these Celsius temperatures to Kelvin temperatures.

   a.   40°C _____

   b.   −30°C _____

Part 4.   The Effect of Temperature on Gases: Charles' Law

1. The volume of a gas is directly proportional to the temperature at constant pressure. This means that when the temperature increases, the volume increases. This relationship between volume and temperature is known as *Charles' Law.*

   If the Kelvin temperature is decreased by one-half, the volume will decrease by ____

   _____ .

6. a. 313 K
   b. 243 K

2. Charles' Law is best expressed with the following equation.

$$\frac{V_1}{T_1} = \frac{V_2}{T_2}$$

T_1 = initial temperature of the gas (K)
V_1 = initial volume of the gas (in mL or L)
T_2 = new temperature of the gas (K)
V_2 = new volume of the gas (in mL or L)

   In the equation for Charles' Law, the new temperature is labeled _____ .

1. one-half

3. In the equation for Charles' Law, the initial volume is labeled _____ .

   a. T_1 b. V_1 c. T_2 d. V_2

2. T_2

4. The Charles' Law equation can be arranged to solve for any of the four quantities listed. Generally, we solve for the new volume.

   Example: Solve for V_2.

   1.   Write the formula.

$$\frac{V_1}{T_1} = \frac{V_2}{T_2}$$

   2.   Multiply both sides by T_2.

$$\frac{V_1 \times T_2}{T_1} = \frac{V_2 \times T_2}{T_2}$$

   3.   Divide like quantities.

$$\frac{V_1 \times T_2}{T_1} = \frac{V_2 \times \cancel{T_2}}{\cancel{T_2}}$$

3. b

4. Rewrite the equation.

$$V_2 = \frac{V_1 T_2}{T_1}$$

Rearrange the Charles' Law equation to solve for T_2.

1.

2.

3.

4.

---

5. Answers to frame 4.

1. $\dfrac{V_1}{T_1} = \dfrac{V_2}{T_2}$

2. $V_1 T_2 = V_2 T_1$

$\dfrac{V_1 T_2}{V_1} = \dfrac{V_2 T_1}{V_1}$

3. $\dfrac{\cancel{V_1} T_2}{\cancel{V_1}} = \dfrac{V_2 T_1}{V_1}$

4. $T_2 = \dfrac{V_2 T_1}{V_1}$

---

6. **We can now solve a Charles' Law problem.**

Example: At 20° C a gas has a volume of 200 *L*. What is the volume of this gas at 127° C?

1. Rearrange the formula.

$$\frac{V_1}{T_1} = \frac{V_2}{T_2} \longrightarrow V_2 T_1 = V_1 T_2$$

$$\frac{V_2 \cancel{T_1}}{\cancel{T_1}} = \frac{V_1 T_2}{T_1}$$

$$V_2 = \frac{V_1 T_2}{T_1}$$

2. List the data. (Remember that all temperatures must be in Kelvin.)

$V_1 = 200\ L$

$T_1 = K = °C + 273$

$\qquad K = 27 + 273 = 300\ K$

$V_2 = ?$

$T_2 = K = °C + 273$

$\qquad K = 127 + 273 = 400\ K$

3. Substitute the data into the formula.

$$V_2 = \frac{(200\ L)(400\ K)}{300\ K}$$

4. Do the calculation.

$$V_2 = \frac{(200\ L)(400\ \cancel{K})}{300\ \cancel{K}} = 267\ L$$

Solve this Charles' Law problem.

At 27°C a gas has a volume of 600 mL. What is the volume of this gas at 107°C?

1.

2.

3.

4.

---

7. Answers to frame 6.

1. $\dfrac{V_1}{T_1} = \dfrac{V_2}{T_2} \longrightarrow V_2 T_1 = V_1 T_2$

   $\dfrac{V_2 \cancel{T_1}}{\cancel{T_1}} = \dfrac{V_1 T_2}{T_1}$

   $V_2 = \dfrac{V_1 T_2}{T_1}$

3. $V_2 = \dfrac{(600 \text{ mL})(380 \text{ K})}{300 \text{ K}}$

4. $V_2 = \dfrac{(600 \text{ mL})(380 \text{ K})}{300 \text{ K}} = 760 \text{ mL}$

2. $V_1 = 600 \text{ mL}$
   $T_1 = K = °C + 273$
   $\qquad K = 27 + 273 = 300 \text{ K}$
   $V_2 = ?$
   $T_2 = K = °C + 273$
   $\qquad K = 107 + 273 = 380 \text{ K}$

---

8. Try this problem.

At 227°C a gas has a volume of 200 mL. What is the volume of this gas at −23°C?

---

9. Answers to frame 8.

1. $\dfrac{V_1}{T_1} = \dfrac{V_2}{T_2} \longrightarrow V_2 T_1 = V_1 T_2$

   $\dfrac{V_2 \cancel{T_1}}{\cancel{T_1}} = \dfrac{V_1 T_2}{T_1}$

   $V_2 = \dfrac{V_1 T_2}{T_1}$

3. $V_2 = \dfrac{(200 \text{ mL})(250 \text{ K})}{500 \text{ K}}$

4. $V_2 = \dfrac{(200 \text{ mL})(250 \text{ K})}{500 \text{ K}} = 100 \text{ mL}$

2. $V_1 = 200 \text{ mL}$
   $T_1 = K = 227 + 273 = 500 \text{ K}$
   $T_2 = K = (−23 + 273) = 250 \text{ K}$

Part 5. The General Gas Law

1. Boyle's Law describes the effect of changing pressure on the volume of a gas when the temperature is held constant. Charles' Law describes the effect of changing temperature on a gas when the pressure is held constant. We will now look at problems in which both the temperature and the pressure change. The formula describing the change in volume when both pressure and temperature change is known as the *general gas law* (or *combined gas law*).

   Boyle's Law deals with what quantities? _____

   a. pressure/temperature
   c. volume/temperature
   b. pressure/volume
   d. volume/temperature/pressure

2. Charles' Law deals with what quantities? _____

   a. pressure/temperature
   c. volume/temperature
   b. pressure/volume
   d. volume/temperature/pressure

   1. b

3. The general gas law deals with what quantities? _____

   a. pressure/temperature
   c. volume/temperature
   b. pressure/volume
   d. volume/temperature/pressure

   2. c

4. The formula for the general gas law is derived by merging the Boyle's Law and Charles' Law formulas.

   3. d

   Boyle's Law: $P_1 V_1 = P_2 V_2$

   Charles' Law: $\dfrac{V_1}{T_1} = \dfrac{V_2}{T_2}$

   general gas law: $\dfrac{P_1 V_1}{T_1} = \dfrac{P_2 V_2}{T_2}$

5. The general-gas-law equation is usually solved for V_2, using the following steps.

   1. Write the formula.

      $$\frac{P_1 V_1}{T_1} = \frac{P_2 V_2}{T_2}$$

   2. Rearrange the equation.

      a. Cross multiply.

      $$\frac{P_1 V_1}{T_1} = \frac{P_2 V_2}{T_2} \longrightarrow P_2 V_2 T_1 = P_1 V_1 T_2$$

      b. Divide each side by the unwanted quantities.

      $$\frac{P_2 V_2 T_1}{P_2 T_1} = \frac{P_1 V_1 T_2}{P_2 T_1}$$

   3. Cancel like quantities.

      $$\frac{\cancel{P_2} V_2 \cancel{T_1}}{\cancel{P_2} \cancel{T_1}} = \frac{P_1 V_1 T_2}{P_2 T_1}$$

4.   Rewrite the equation.

$$V_2 = \frac{P_1 V_1 T_2}{P_2 T_1}$$

Solve the general-gas-law formula for P_2.

1.

2.

3.

4.

---

6.   Answers to frame 5.

1. $\dfrac{P_1 V_1}{T_1} = \dfrac{P_2 V_2}{T_2}$

2. $P_2 V_2 T_1 = P_1 V_1 T_2$

   $\dfrac{P_2 V_2 T_1}{V_2 T_1} = \dfrac{P_1 V_1 T_2}{V_2 T_1}$

3. $\dfrac{P_2 \cancel{V_2} \cancel{T_1}}{\cancel{V_2} \cancel{T_1}} = \dfrac{P_1 V_1 T_2}{V_2 T_1}$

4. $P_2 = \dfrac{P_1 V_1 T_2}{V_2 T_1}$

---

7.   We can now solve general-gas-law problems.

Example: A gas has a volume of 400 mL and 4.00 atm of pressure at −73°C. What will its volume be at 27°C and 2.00 atm of pressure?

1. $\dfrac{P_1 V_1}{T_1} = \dfrac{P_2 V_2}{T_2} \longrightarrow P_2 V_2 T_1 = P_1 V_1 T_2$

   $\dfrac{\cancel{P_2} V_2 \cancel{T_1}}{\cancel{P_2} \cancel{T_1}} = \dfrac{P_1 V_1 T_2}{P_2 T_1}$

   $V_2 = \dfrac{P_1 V_1 T_2}{P_2 T_1}$

2. $P_1 = 4.00$ atm
   $V_1 = 400$ mL
   $T_1 = (-73 + 273) = 200$ K
   $P_2 = 2.00$ atm
   $V_2 = ?$
   $T_2 = 27 + 273 = 300$ K

3. $V_2 = \dfrac{(4.00 \text{ atm})(400 \text{ mL})(300\text{K})}{(2.00 \text{ atm})(200 \text{ K})}$

4. $V_2 = \dfrac{(4.00 \cancel{\text{ atm}})(400 \text{ mL})(300\cancel{\text{K}})}{(2.00 \cancel{\text{ atm}})(200 \cancel{\text{ K}})}$

   $V_2 = 1200$ mL

---

8. Try this problem.

A gas has a volume of 8.0 L at 127°C and 300 torr. What would the volume of the gas be at −23°C and 600 torr?

9. Answers to frame 8.

1. $\dfrac{P_1V_1}{T_1} = \dfrac{P_2V_2}{T_2} \longrightarrow V_2 = \dfrac{P_1V_1T_2}{T_1P_2}$

2. $P_1 = 300$ torr
   $V_1 = 8.0\ L$
   $T_1 = 127 + 273 = 400\ K$
   $P_2 = 600$ torr
   $V_2 = ?$
   $T_2 = (-23 + 273) = 250\ K$

3. $V_2 = \dfrac{(300\ \text{torr})(8.0\ L)(250K)}{(400\ K)(600\ \text{torr})}$

4. $V_2 = \dfrac{(300\ \cancel{\text{torr}})(8.0\ L)(250\cancel{K})}{(400\ \cancel{K})(600\ \cancel{\text{torr}})}$

   $V_2 = 2.5\ L$

10. A gas occupies a volume of 200 mL at 600 torr. What will its volume be at 800 torr?

11. Answers to frame 10.

1. $P_1V_1 = P_2V_2 \longrightarrow V_2 = \dfrac{P_1V_1}{P_2}$

2. $P_1 = 600$ torr
   $V_1 = 200$ mL
   $P_2 = 800$ torr
   $V_2 = ?$

3. $V_2 = \dfrac{(600\ \cancel{\text{torr}})(200\ \text{mL})}{800\ \cancel{\text{torr}}}$

4. $V_2 = 150$ mL

12. A gas occupies 8.0 L at 4.0 atm of pressure. What will its volume be at 10 atm of pressure?

13. Answers to frame 12.

1. $P_1V_1 = P_2V_2 \longrightarrow V_2 = \dfrac{P_1V_1}{P_2}$

2. $P_1 = 4.0$ atm
   $V_1 = 8.0\ L$
   $P_2 = 10$ atm
   $V_2 = ?$

3. $V_2 = \dfrac{(4.0\ \cancel{atm})(8.0\ L)}{10\ \cancel{atm}}$

4. $V_2 = 3.2\ L$

---

14. A gas has a volume of 100 mL at 27°C. What volume will it occupy at −73°C?

---

15. Answers to frame 14.

1. $\dfrac{V_1}{T_1} = \dfrac{V_2}{T_2} \longrightarrow V_2 = \dfrac{V_1 T_2}{T_1}$

2. $V_1 = 100$ mL
   $T_1 = 27 + 273 = 300\,K$
   $V_2 = ?$
   $T_2 = (-73 + 273) = 200\,K$

3. $V_2 = \dfrac{(100\ mL)(200\,\cancel{K})}{300\ \cancel{K}}$

4. $V_2 = 66.7$ mL

---

16. A gas has a volume of 10 L at −23°C. What volume will it occupy at 227°C?

---

17. Answers to frame 16.

1. $\dfrac{V_1}{T_1} = \dfrac{V_2}{T_2} \longrightarrow V_2 = \dfrac{V_1 T_2}{T_1}$

2. $V_1 = 10\ L$
   $T_1 = (-23 + 273) = 250\,K$
   $V_2 = ?$
   $T_2 = 227 + 273 = 500\,K$

3. $V_2 = \dfrac{(10\ L)(500\ \cancel{K})}{250\,\cancel{K}}$

4. $V_2 = 20\ L$

18. A gas has a volume of $15\,L$ at $27°C$ at 4.0 atm of pressure. What volume will it occupy at $127°C$ and 2.0 atm of pressure?

19. Answers to frame 18.

1. $\dfrac{P_1 V_1}{T_1} = \dfrac{P_2 V_2}{T_2} \longrightarrow V_2 = \dfrac{P_1 V_1 T_2}{T_1 P_2}$

2. $P_1 = 4.0$ atm
   $V_1 = 15\,L$
   $T_1 = 27°C + 273 = 300\,K$
   $P_2 = 2.0$ atm
   $V_2 = ?$
   $T_2 = 127 + 273 = 400\,K$

3. $V_2 = \dfrac{(4.0\ \text{atm})(15\,L)(400\ \text{K})}{(2.0\ \text{atm})(300\ \text{K})}$

4. $V_2 = 40\,L$

20. A gas has a volume of 750 mL at $-23°C$ and 1000 torr of pressure. What volume will it occupy at $27°C$ and 500 torr of pressure?

21. Answers to frame 20.

1. $\dfrac{P_1 V_1}{T_1} = \dfrac{P_2 V_2}{T_2} \longrightarrow V_2 = \dfrac{P_1 V_1 T_2}{T_1 P_2}$

2. $P_1 = 1000$ torr
   $V_1 = 750$ mL
   $T_1 = (-23 + 273) = 250\,K$
   $P_2 = 500$ torr
   $V_2 = ?$
   $T_2 = 27 + 273 = 300\,K$

3. $V_2 = \dfrac{(1000\ \text{torr})(750\ \text{mL})(300\ \text{K})}{(500\ \text{torr})(250\ \text{K})}$

4. $V_2 = 1800$ mL

Part 6. The Ideal-Gas Equation

1. You have learned how the volume of a gas can vary with temperature and pressure. The number of moles of molecules that a given amount of gas contains will also have an effect on the volume and pressure of the gas. Two moles of a gas will occupy twice the volume that one mole would under the same temperature and pressure. This relationship is best described in the ideal-gas equation: $PV = nRT$.

P = pressure in atm
V = volume in L
n = number of moles of gas
R = gas constant: 0.0821 L-atm/mole-K (which is read as liter-atmosphere per mole-K)
T = temperature in K

2. When the number of moles of a gas are decreased but the temperature and pressure remain the same, the volume of the gas will _____ .

3. State what units of measurement the following quantities must be in when using the ideal-gas equation.

   a. Pressure must be in _____ .

   b. Temperature must be in _____ .

   c. Volume must be in _____ .

2. decrease

4. Give the numerical value and units of the gas constant. _____

3. a. atmospheres
   b. degrees Kelvin
   c. liters

5. To make it easier to compare volumes of gases, a standard temperature and pressure (STP) have been selected. The standard temperature is 0°C (273K) and the standard pressure is 1 atm (760 torr).

   a. Standard temperature and pressure are abbreviated as _____ .

   b. The standard temperature is _____ °C.

   c. The standard pressure is _____ atm.

4. 0.0821 L-atm/mole-K

6. Let's look at an example of an ideal-gas-equation problem.

   Example: What volume does 1.00 mole of oxygen occupy at 0°C and 1.00 atm of pressure?

   1. Rearrange the equation to solve for V by dividing each side by the unwanted quantity.

$$PV = nRT \qquad \frac{\cancel{R}V}{\cancel{R}} = \frac{nRT}{P} \qquad V = \frac{nRT}{P}$$

   2. List the data.

P = 1.00 atm R = 0.0821 L-atm/mole-K
V = ? T = 0°C + 273 = 273K
n = 1.00 mole

5. a. STP
   b. 0
   c. 1

3. Substitute the data into the equation.

$$V = \frac{(1.00 \text{ mole})(0.0821 \ L\text{-atm})(273K)}{(1.00 \text{ atm})(\text{mole-K})}$$

Note how the units of the gas constant are handled here. The L-atm units are placed in the numerator and the mole-K units are placed in the denominator.

4. Do the calculation.

a. Cancel like units.

$$V = \frac{(1.00 \ \cancel{\text{mole}})(0.0821 \ L\text{-}\cancel{\text{atm}})(273\cancel{K})}{(1.00 \ \cancel{\text{atm}})(\cancel{\text{mole-K}})}$$

Note that only the unit *liter* is left.

b. Do the arithmetic.

$$V = \frac{1.00)(0.0821 \ L)(273)}{1.00} \qquad V = 22.4 \ L$$

This volume is known as the *molar volume* (or gram-molecular volume) of a gas at STP. One mole of any gas will occupy 22.4 L at STP.

---

7. At STP, one mole of any gas will occupy _____ L.

---

8. What volume does 2.50 moles of nitrogen gas occupy at STP?

7. 22.4 L

---

9. Answers to frame 8.

1. $PV = nRT \qquad V = \dfrac{nRT}{P}$

2. $P = 1.00$ atm $\qquad R = 0.821 \ L$-atm/mole-K
   $V = ?$ $\qquad T = 0°C + 273 = 273 \ K$
   $n = 2.50$ moles

3. $V = \dfrac{(2.50 \text{ moles})(0.0821 \ L\text{-atm})(273K)}{(1.00 \text{ atm})(\text{mole-K})}$

4. $V = \dfrac{(2.50 \ \cancel{\text{moles}})(0.0821 \ L\text{-}\cancel{\text{atm}})(273\cancel{K})}{(1.00 \ \cancel{\text{atm}})(\cancel{\text{mole-K}})}$

   $V = 56.0 \ L$

---

10. At what pressure would 2.50 moles of chlorine at 20°C occupy 20.5 L?

11. Answers to frame 10.

   1. $PV = nRT$ $P = \dfrac{nRT}{V}$

   2. $P = ?$ $R = 0.0821$ L-atm/mole-K
        $V = 20.5\ L$ $T = 20°C + 273 = 293\,K$
        $n = 2.50$ moles

   3. $P = \dfrac{(2.50\ \cancel{moles})(0.0821\ \cancel{L}\text{-atm})(293\cancel{K})}{(20.5\ \cancel{L})(\cancel{\text{mole-K}})}$

   4. $P = 2.93$ atm

---

12. Because you know how to calculate molecular mass and find the number of moles that a given weight of a compound represents, you can solve a variation of the ideal-gas equation.

   Example: What volume does 80.0 g of sulfur dioxide (SO_2) gas occupy at 15°C and 2.80 atm of pressure?

   1. $PV = nRT$ $V = \dfrac{nRT}{P}$

   2. $P = 2.80$ atm $R = 0.0821$ L-atm/mole-K
        $V = ?$ $T = 15°C + 273 = 288\,K$

        $n =$ a. number of moles $= \dfrac{\text{weight in grams}}{\text{weight of 1 mole}}$

                b. weight in grams $= 80.0$ g
                   weight of 1 mole of S $= 32.1 \times 1 = 32.1$ g
                   weight of 1 mole of O $= 16.0 \times 2 = \underline{32.0}$ g
                                         64.1 g

                c. number of moles $= \dfrac{80.0\ \text{g}}{64.1\ \text{g}}$

                d. number of moles $= 1.25$ moles

   3. $V = \dfrac{(1.25\ \cancel{moles})(0.0821\ L\text{-}\cancel{atm})(288\cancel{K})}{(2.80\ \cancel{atm})(\cancel{\text{mole-K}})}$

   4. $V = 10.6\ L$

---

13. What volume would 35.0 g of methane gas (CH_4) occupy at 20°C and 1.50 atm of pressure?

14. Answers to frame 13.

   1. $PV = nRT$ $V = \dfrac{nRT}{P}$

   2. $P = 1.50$ atm $R = 0.0821$ L-atm/mole-K
       $V = ?$ $T = 20°C + 273 = 293$ K

       $n = $ a. number of moles $= \dfrac{\text{weight in grams}}{\text{weight of 1 mole}}$

           b. weight in grams $= 35.0$ g
              weight of 1 mole of C $= 12.0 \times 1 = 12.0$ g
              weight of 1 mole of H $= 1.0 \times 4 = \underline{\quad 4.0 \text{ g}}$
                                    16.0 g

           c. number of moles $= \dfrac{35.0 \text{ g}}{16.0 \text{ g}}$

           d. number of moles $= 2.19$ moles

   3. $V = \dfrac{(2.19 \text{ moles})(0.0821 \text{ } L\text{-atm})(293\text{K})}{(1.50 \text{ atm})(\text{mole-K})}$

   4. $V = 35.1$ L

---

15. It is often necessary to find the density of a gas at specific temperatures and pressures. This is accomplished in two parts. First solve the ideal-gas equation and then use the answer obtained to solve for the density.

   Example: What is the density of 1.00 mole of CO_2 gas at 15°C and 0.750 atm?

   a. Solve the ideal-gas equation to find the volume.

       1. $PV = nRT$ $V = \dfrac{nRT}{P}$

       2. $P = 0.750$ atm $R = 0.0821$ L-atm/mole-K
            $V = ?$ $T = 15°C + 273 = 288$ K
            $n = 1.00$ mole

       3. $V = \dfrac{(1.00 \text{ mole})(0.0821 \text{ } L\text{-atm})(288\text{K})}{(0.750 \text{ atm})(\text{mole-K})}$

       4. $V = 35.1$ L

   b. Solve for the density of the gas.

       1. density $= \dfrac{\text{mass in grams}}{\text{volume in liters}}$ The mass must be in grams. Use the formula that converts moles to grams.

       2. a. mass in grams $= $ (number of moles)(weight of 1 mole)
           b. number of moles $= 1.00$ mole
              weight of 1 mole of C $= 12.0 \times 1 = 12.0$ g
              weight of 1 mole of O $= 16.0 \times 2 = \underline{32.0 \text{ g}}$
                                  44.0 g/mole
           c. mass in grams $= (1.00 \text{ mole})(44.0 \text{ g/mole})$
           d. mass in grams $= 44.0$ g
             volume $= 31.5$ L

       3. density $= \dfrac{44.0 \text{ g}}{31.5 \text{ } L}$

       4. density $= 1.40$ g/L

16. What is the density of 2.52 moles of SO_3 gas at 10°C and 1.25 atm?

---

17. Answers to frame 16.

a. 1. $PV = nRT$ $V = \dfrac{nRT}{P}$

   2. $P = 1.25$ atm $R = 0.0821$ L-atm/mole-K
        $V = ?$ $T = 10°C + 273 = 283$ K
        $n = 2.52$ moles

   3. $V = \dfrac{(2.52 \text{ moles})(0.0821 \text{ } L\text{-atm})(283 \text{ K})}{(1.25 \text{ atm})(\text{mole-K})}$

   4. $V = 46.8$ L

b. 1.   density $= \dfrac{\text{mass in grams}}{\text{volume in liters}}$

   2.   a. mass in grams $=$ (number of moles)(weight of 1 mole)
        b. number of moles $= 2.52$ moles
           weight of 1 mole of S $= 32.1 \times 1 = 32.1$ g
           weight of 1 mole of O $= 16.0 \times 3 = \underline{48.0 \text{ g}}$
                                                    80.1 g
        c. mass $= (2.52 \text{ moles})(80.1 \text{ g/mole})$
        d. mass $= 202$ g

   3.   density $= \dfrac{202 \text{ g}}{46.8 \text{ } L}$

   4.   density $= 4.32$ g/L

---

18. What is the volume of 2.60 moles of hydrogen gas at STP?

---

19. Answers to frame 18.

   1. $V = \dfrac{nRT}{P}$

   2. $P = 1.00$ atm $R = 0.0821$ L-atm/mole-K
        $V = ?$ $T = 0°C + 273 = 273$ K
        $n = 2.60$ moles

   3. $V = \dfrac{(2.60 \text{ moles})(0.0821 \text{ } L\text{-atm})(273 \text{ K})}{(1.00 \text{ atm})(\text{mole-K})}$

   4. $V = 58.3$ L

20. What volume does 15.0 g of NO_2 gas occupy at $12°C$ and 0.520 atm?

21. Answers to frame 20.

   1. $V = \dfrac{nRT}{P}$

   2. P = 0.0520 atm R = 0.0821 L-atm/mole-K
       V = ? T = $12°C$ + 273 = 285 K

      n = a. number of moles = $\dfrac{\text{weight in grams}}{\text{weight of 1 mole}}$

           b. weight in grams = 15.0 g
              weight of 1 mole of N = 14.0 $\times$ 1 = 14.0 g
              weight of 1 mole of O = 16.0 $\times$ 2 = $\underline{32.0\ \text{g}}$
                                         46.0 g/mole

           c. number of moles = $\dfrac{15.0\ \text{g}}{46.0\ \text{g/mole}}$

           d. number of moles = 0.326 mole

   3. $V = \dfrac{(0.326\ \text{mole})(0.0821\ L\text{-atm})(285\,\text{K})}{(0.520\ \text{atm})(\text{mole-K})}$

   4. V = 14.7 L

22. What is the density of 1.18 moles of CH_4 gas at $15°C$ and 2.42 atm of pressure?

23. Answers to frame 22.

  a. 1. $V = \dfrac{nRT}{P}$

     2. P = 2.42 atm R = 0.0821 L-atm/mole-K
        V = ? T = $15°C$ + 273 = 288 K
        n = 1.18 moles

     3. $V = \dfrac{(1.18\ \text{moles})(0.0821\ L\text{-atm})(288\,\text{K})}{(2.42\ \text{atm})(\text{mole-K})}$

     4. V = 11.5 L

(Answers to frame 22. Continued)

b. 1.   density $= \dfrac{\text{mass in grams}}{\text{volume}}$

2.   a. mass = (number of moles)(weight of 1 mole)
   b. number of moles = 1.18 moles
      weight of 1 mole of C = 12.0 X 1 = 12.0 g
      weight of 1 mole of H = 1.0 X 4 = 4.0 g

      16.0 g/mole

   c. mass = (1.18 moles)(16.0 g/mole)
   d. mass = 18.9 g

3.   density $= \dfrac{18.9\ g}{11.5\ L}$

4.   density = 1.64 g/L

Part 7. Stoichiometry Involving Gases

1. A variety of stoichiometric calculations involving gases can be performed using the following conversion factors.

a. Number of moles $= \dfrac{\text{wt in g}}{\text{wt/mole}}$

or wt in g = (number of moles)(wt/mole)

b. Number of moles $= \dfrac{\text{liters of gas}}{22.4\ L/\text{mole (at STP)}}$

or liters of gas = (number of moles) (22.4L/mole (at STP))

Example: How many liters of O_2 at STP can be formed from 0.300 mole of $KClO_3$ in the following reaction? (First solve the mole/mole problem, and then change the answer to liters.)

Part A.   Step 1.   2$KClO_3$ $\longrightarrow$ KCl + 3O_2
                    2 moles   3 moles

Step 2.   mole ratio $= \dfrac{3\text{ moles }O_2}{2\text{ moles }KClO_3}$

Step 3.   0.300 mole $KClO_3$ X $\dfrac{3\text{ moles }O_2}{2\text{ moles }KClO_3}$ = 0.450 moles O_2

Part B.   Change moles to liters.

Step 1.   liters of gas at STP = (number of moles of gas) (22.4 L/mole)

Step 2.   Number of moles of gas = 0.450 moles O_2

Step 3.   Number of liters of gas at STP = 0.450 moles X $\dfrac{24.4\ L}{1\text{ mole}}$

Step 4.   Number of liters of gas at STP = 10.1 L

or 0.300 mole $KClO_3$ X $\dfrac{3\text{ moles }O_2}{2\text{ moles }KClO_3}$ X $\dfrac{22.4 LO_2}{1\text{ mole }O_2}$ =10.1 L

Do these problems.

a. How many liters of O_2 gas at STP will be produced by 0.225 mole of HgO in the following reaction?

2HgO $\longrightarrow$ 2Hg + O_2

b. How many liters of CO_2 gas at STP will be produced by 1.50 moles of C_2H_6 in the following reaction?

$$2C_2H_6 + 7O_2 \longrightarrow 4CO_2 + 6H_2O$$

---

2. Answers to frame 1.

a. 0.225 mole HgO $\times \dfrac{1 \text{ mole } O_2}{2 \text{ moles HgO}} \times \dfrac{22.4 \text{ } L \text{ } O_2}{1 \text{ mole } O_2} = 2.52 \text{ } L$

b. 1.50 moles $C_2H_6 \times \dfrac{4 \text{ moles } CO_2}{2 \text{ moles } C_2H_6} \times \dfrac{22.4 \text{ } L \text{ } CO_2}{1 \text{ mole } CO_2} = 67.2 \text{ } L$

---

3. In another type of problem, you will be given the starting substance in grams and be asked to find the volume of gas produced at STP.

Example: How many liters of H_2 gas at STP will 3.44 g of Al produce in the following reaction?

$$2Al + 6HCl \longrightarrow 2AlCl_3 + 3H_2$$

Step 1.   Change grams of Al to moles.

$$\text{moles of Al} = \frac{3.44 \text{ g}}{27.0 \text{ g/mole}} = 0.127 \text{ mole}$$

Step 2.   Multiply the moles of starting substance (Al) by the mole ratio to get the moles of H_2 produced.

$$0.127 \text{ mole Al} \times \frac{3 \text{ moles } H_2}{2 \text{ moles Al}} = 0.191 \text{ mole } H_2$$

Step 3.   Change moles of H_2 to liters at STP.

liters = (number of moles)$(22.4 \text{ } L/\text{mole})$
liters = 0.191 mole $H_2 \times 22.4 \text{ } L/\text{mole } H_2$
liters = $4.28 \text{ } L$ at STP

The conversion formula is:

$$3.44 \text{ g Al} \times \frac{1 \text{ mole Al}}{27.0 \text{ g Al}} \times \frac{3 \text{ moles } H_2}{2 \text{ moles Al}} \times \frac{22.4 \text{ } L}{1 \text{ mole } H_2} = 4.28 \text{ } L$$

Try these problems.

a. How many liters of CO_2 gas will be produced at STP from 22.5 g of C_2H_6 in the following reaction?

$$2C_2H_6 + 7O_2 \longrightarrow 4CO_2 + 6H_2O$$

b. How many liters of O_2 will be needed to react with the 22.5 g of C_2H_6 in the above problem?

4. Answers to frame 3.

a. $22.5 \text{ g } C_2H_6 \times \dfrac{1 \text{ mole } C_2H_6}{30.0 \text{ g } C_2H_6} \times \dfrac{4 \text{ moles } CO_2}{2 \text{ moles } C_2H_6} \times \dfrac{22.4 \text{ L}}{1 \text{ mole } CO_2} = 33.6 \text{ L}$

b. $22.5 \text{ g } C_2H_6 \times \dfrac{1 \text{ mole } C_2H_6}{30.0 \text{ g } C_2H_6} \times \dfrac{7 \text{ moles } O_2}{2 \text{ moles } C_2H_6} \times \dfrac{22.4 \text{ L}}{1 \text{ mole } O_2} = 58.8 \text{ L}$

5. Reactions involving gases do not always occur at STP. The ideal gas equation is used to find the volume of a gas at a temperature and pressure other than STP.

Example: What volume of O_2 will be produced at 20.0°C and 0.850 atm from 30.0 g of $KClO_3$ in the following reaction?

$$2KClO_3 \longrightarrow 2KCl + 3O_2$$

Step 1.   Change grams of starting substance to moles.

$$\text{moles } KClO_3 = \frac{30.0 \text{ g}}{122.6 \text{ g/mole}} = 0.245 \text{ mole}$$

Step 2.   Multiply the moles of $KClO_3$ by the mole ratio to get the moles of O_2 produced.

$$0.245 \text{ mole } KClO_3 \times \frac{3 \text{ moles } O_2}{2 \text{ moles } KClO_3} = 0.368 \text{ mole } O_2$$

Step 3.   Use the ideal-gas equation to find the volume of the gas at 20.0°C and 0.850 atm.

$$V = \frac{nRT}{P}$$

$n = 0.368 \text{ mole } O_2$ $\qquad\qquad$ $T = 20.0°C + 273 = 293 \text{ K}$

$R = 0.0821 \text{ L-atm/mole-K}$ $\qquad\qquad$ $P = 0.850 \text{ atm}$

$$V = \frac{(0.368 \text{ mole})(0.0821 \text{ L-atm})(293 \text{ K})}{(0.850 \text{ atm})(\text{mole-K})}$$

$V = 10.4 \text{ L}$

Try these problems.

a. What volume of CO_2 will be produced at 30.0°C and 0.950 atm from 20.0 g of C_2H_6 in the following reaction?

$$2C_2H_6 + 7O_2 \longrightarrow 4CO_2 + 6H_2O$$

b. What volume of H_2 will be produced at 25°C and 1.10 atm from 40.0 g of Al in the following reaction?

$$2Al + 6HCl \longrightarrow 2AlCl_3 + 3H_2$$

6. Answers to frame 5.

a. $20.0 \text{ g C}_2\text{H}_6 \times \dfrac{1 \text{ mole C}_2\text{H}_6}{30.0 \text{ g C}_2\text{H}_6} \times \dfrac{4 \text{ moles CO}_2}{2 \text{ moles C}_2\text{H}_6} = 1.33 \text{ moles CO}_2 \text{ produced}$

$V = \dfrac{(1.33 \text{ moles CO}_2(0.0821 \text{ } L \text{ atm})(303\text{K})}{(0.950 \text{ atm})(\text{mole K})} = 34.8 \text{ } L$

b. $40.0 \text{ g Al} \times \dfrac{1 \text{ mole Al}}{27.0 \text{ g Al}} \times \dfrac{3 \text{ moles H}_2}{2 \text{ moles Al}} = 2.22 \text{ moles H}_2 \text{ produced}$

$V = \dfrac{(2.22 \text{ moles H}_2)(0.0821 \text{ } L\text{-atm})(298\text{K})}{(1.10 \text{ atm})(\text{mole-K})} = 49.4 \text{ } L$

7. Since a mole of any gas equals 22.4 liters at STP, we can substitute the volumes of gases for moles of gases in reactions where all substances are gases.

$$3\text{H}_2 + \text{N}_2 \longrightarrow 2\text{NH}_3$$

| 3 moles | 1 mole | 2 moles |
| 3 volumes | 1 volume | 2 volumes |

Using the volumes of the gases greatly simplifies the calculations. This type of problem is known as a volume/volume calculation.

Example: What volume of H_2 will react with 10 liters of N_2 in the following reaction?

$$3\text{H}_2 + \text{N}_2 \longrightarrow 2\text{NH}_3$$

| 3 volumes | 1 volume | 2 volumes |

Multiply the liters or milliliters of starting substance (10 L of N_2) by a volume ratio instead of a mole ratio.

$10 \text{ L N}_2 \times \dfrac{3 \text{ volumes H}_2}{1 \text{ volume N}_2} = 30 \text{ } L \text{ H}_2$

Try these problems.

a. What volume of NH_3 will be formed from 20 liters of N_2 in the reaction above?

b. What volume of NH_3 will be formed from 600 mL of H_2 in the reaction named?

8. Answers to frame 7.

a. $20 \text{ } L \text{ N}_2 \times \dfrac{2 \text{ volumes NH}_3}{1 \text{ volume N}_2} = 40 \text{ } L \text{ NH}_3$

b. $600 \text{ mL H}_2 \times \dfrac{2 \text{ volumes N}_2}{3 \text{ volumes H}_2} = 400 \text{ mL NH}_3$

9. How many liters of H_2 gas at STP will be produced by 0.300 mole of Al in the following reaction?

$$2Al + 6HCl \longrightarrow 2AlCl_3 + 3H_2$$

10. Answers to frame 9.

$$0.300 \text{ mole Al} \times \frac{3 \text{ moles } H_2}{2 \text{ moles Al}} \times \frac{22.4 \text{ } L}{1 \text{ mole } H_2} = 10.1 \text{ L}$$

11. How many liters of CO_2 gas will be produced at STP by 50.5 g of $NaHCO_3$ in the following reaction?

$$2NaHCO_3 \longrightarrow Na_2CO_3 + CO_2 + H_2O$$

12. Answers to frame 11.

$$50.0 \text{ g } NaHCO_3 \times \frac{1 \text{ mole } NaHCO_3}{84.0 \text{ g } NaHCO_3} \times \frac{1 \text{ mole } CO_2}{2 \text{ moles } NaHCO_3} \times \frac{22.4 \text{ } L}{1 \text{ mole } CO_2} = 6.73 \text{ } L$$

13. How many liters of NH_3 gas will be produced from 10.7 g of N_2 at STP in the following reaction?

$$N_2 + 3H_2 \longrightarrow 2NH_3$$

14. Answers to frame 13.

$$10.7 \text{ g } N_2 \times \frac{1 \text{ mole } N_2}{28.0 \text{ g } N_2} \times \frac{2 \text{ moles } NH_3}{1 \text{ mole } N_2} \times \frac{22.4 \text{ L}}{1 \text{ mole } NH_3} = 17.1 \text{ } L$$

15. What volume of CO_2 will be produced at 22.0°C and 0.950 atm from 50.0 g of C_2H_5OH in the following reaction?

$$C_2H_5OH + 3O_2 \longrightarrow 2CO_2 + 3H_2O$$

16. Answers to frame 15.

$$50.0 \text{ g } C_2H_5OH \times \frac{1 \text{ mole } C_2H_5OH}{46.0 \text{ g } C_2H_5OH} \times \frac{2 \text{ moles } CO_2}{1 \text{ mole } C_2H_5OH} = 2.17 \text{ moles produced}$$

$$V = \frac{(2.17 \text{ moles})(0.0821 \text{ } L\text{-atm})(295 \text{ K})}{(0.950 \text{ atm})(\text{mole-K})} = 55.3 \text{ } L$$

---

17. What volume of H_2 will be produced at $30.0°C$ and 1.10 atm from 12.0 g of Na in the following reaction?

$$2Na + 2H_2O \longrightarrow 2NaOH + H_2$$

---

18. Answers to frame 17.

$$12.0 \text{ g Na } \times \frac{1 \text{ mole Na}}{23.0 \text{ g Na}} \times \frac{1 \text{ mole } H_2}{2 \text{ moles Na}} = 0.261 \text{ mole } H_2 \text{ produced}$$

$$V = \frac{(0.261 \text{ moles})(0.0821 \text{ } L\text{-atm})(303 \text{ K})}{(1.10 \text{ atm})(\text{mole-K})} = 5.90 \text{ } L$$

---

19. What volume of CO_2 will be produced from 10 liters of O_2 in the following reaction?

$$CH_4 + 2O_2 \longrightarrow CO_2 + 2H_2O$$

---

20. Answers to frame 19.

$$10 \text{ } L \text{ } O_2 \times \frac{1 \text{ volume } CO_2}{2 \text{ volumes } O_2} = 5.0 \text{ } L \text{ } CO_2$$

---

21. How many liters of O_2 are needed to react with 4.50 liters of CO in the following reaction?

$$2CO + O_2 \longrightarrow 2CO_2$$

---

22. Answers to frame 21.

$$4.50 \text{ } L \text{ } CO \times \frac{1 \text{ volume } O_2}{2 \text{ volumes } CO} = 2.25 \text{ } L \text{ } O_2$$

Part 8. Collecting Gases Over Water

1. A commonly performed laboratory experiment is collecting a gas over water. The ideal-gas equation can be used for this purpose.

In the laboratory:

a. Pressure will be read in torrs and must be converted to atmospheres. The water evaporates and causes a slight pressure that must be subtracted from the total pressure. The amount of this pressure is given in a chart or must be stated in the problem.

b. The volume is usually obtained in milliliters and must be converted to liters.

c. The temperature will be obtained in Celsius degrees and must be converted to Kelvin degrees.

Example: 257 mL of N_2 gas are collected over water at 20°C and 800 torr. The amount of gas weighs 0.300 g and the vapor pressure of water at 20°C is 17.0 torr. What is the molecular weight of N_2?

Step 1. Rearrange the ideal-gas equation to solve for molecular weight.

$$PV = nRT$$

$$n = \frac{PV}{RT} \quad (n = \text{number of moles})$$

A substitution is made for n to obtain the weight/mole. The weight/mole equals the molecular weight (MW).

$$n = \frac{\text{wt in g}}{\text{wt/mole (MW)}}$$

$$\frac{\text{wt in g}}{\text{MW}} = \frac{PV}{RT}$$

Solving for MW

$$MW = \frac{(\text{wt in g})(T)(R)}{(P)(V)}$$

Step 2. List and convert your data to the desired units.

a. wt in g = 0.300 g (no conversion needed)

b. pressure = total pressure − water vapor pressure

$$800 \text{ torr} - 17.0 \text{ torr} = 783 \text{ torr}$$

$$783 \text{ torr} \times \frac{1 \text{ atm}}{760 \text{ torr}} = 1.03 \text{ atm}$$

c. volume = $257 \text{ mL} \times \frac{1 \text{ liter}}{1000 \text{ mL}} = 0.257 L$

d. temperature = 20°C + 273 = 293 K

e. R = 0.082 L-atm/mole-K

Step 3. Substitute into the equation.

$$MW = \frac{(0.300 \text{ g})(293 \text{ K})(0.0821 \text{ } L\text{-atm})}{(1.03 \text{ atm})(0.257 \text{ } L)(\text{mole-K})}$$

Step 4. Solve

$$MW = 28.0 \text{ g/mole}$$

Do this problem.

261 mL of CO_2 gas are collected over water at 27°C and 760 torr. The amount of gas weighs 0.449 g and the vapor pressure of water at 27°C is 27.0 torr. What is the molecular weight of CO_2?

---

2. Answer to frame 1.
   1. MW = gRT/PV

   2. g = 0.449 g

      P = 760 torr − 27.0 torr = 733 torr

      733 torr × 1 atm/760 torr = 0.964 atm

      V = 261 mL × 1 L/1000 mL = 0.261 L

      T = 27°C + 273 = 300 K

      R = 0.0821 L-atm/mole-K

   3. MW = $\dfrac{(0.449 \text{ g})(0.0821 \text{ } L\text{-atm})(300 \text{ K})}{(0.964 \text{ atm})(0.261 \text{ } L)(\text{mole-K})}$

   4. MW = 44.0 g/mole

---

3. 273 mL of NO_2 gas are collected over water at 21°C and 750 torr. The amount of gas collected weighs 0.500 g and the vapor pressure of water at 21°C is 19.0 torr. What is the molecular weight of NO_2?

---

4. Answer to frame 3.

   MW = $\dfrac{(0.500 \text{ g})(0.0821 \text{ } L\text{-atm})(294 \text{ K})}{(0.962 \text{ atm})(0.273 \text{ } L)(\text{mole-K})}$

   MW = 46.0 g/mole

NAME _____

THE GAS LAWS
EVALUATION TEST 1
PART 1–PART 2

1. Average atmospheric pressure at sea level is (a) _____ torr or (b) _____ atm.

2. 380 torr is _____ atm.

3. 2 atm are _____ torr.

4. When the pressure on a gas increases, will the volume increase or decrease?

5. If the pressure on a gas is decreased by one-half, how large will the volume be?

Show all steps in solving the problems below.

6. Solve the Boyle's Law equation for V_2.

1. _____

2. _____

3. _____

4. _____

5. _____

6. _____

7. _____

8. _____

9. _____

7. 600 mL of a gas are under a pressure of 8 atm. What would the volume of the gas be at 2 atm?

8. 400 mL of a gas are under a pressure of 800 torr. What would the volume of the gas be at a pressure of 1000 torr?

9. 4 L of a gas are under a pressure of 6 atm. What would the volume of the gas be at 2 atm?

NAME _____

THE GAS LAWS
EVALUATION TEST 2
PART 3-PART 4

1.   The boiling point of water on the Kelvin scale is _____ .

2.   0 K is also known as _____ .

3.   When the temperature of a gas decreases, does the volume increase or decrease?

4.   If the Kelvin temperature of a gas is doubled, the volume of the gas will increase by

     _____ .

Show all steps in the problems below.

5.   40°C is what temperature on the Kelvin scale?

6.   −20°C is what temperature on the Kelvin scale?

7.   Solve the Charles' Law equation for V_2.

8.   At 27°C a gas has a volume of 6.0 *L.* What is the volume of this gas at 150°C?

9.   At 225°C a gas has a volume of 400 mL. What is the volume of this gas at 127°C?

10. At 210°C a gas has a volume of 8.0 *L.* What is the volume of this gas at −23°C?

1. _____

2. _____

3. _____

4. _____

5. _____

6. _____

7. _____

8. _____

9. _____

10. _____

NAME _____

THE GAS LAWS
EVALUATION TEST 3
PART 5

1. Boyle's Law deals with what quantities?

    a. pressure/temperature   b. pressure/volume
    c. volume/temperature   d. volume/temperature/pressure

2. Charles' Law deals with what quantities?

    a. pressure/temperature   b. pressure/volume
    c. volume/temperature   d. volume/temperature/pressure

3. Write Charles' Law in equation form.

4. Write the general gas law in equation form.

Show all steps for the following problems.

5. Solve the general gas law for V_2.

6. A gas has a volume of 800 mL at −23° and 300 torr. What would the volume of the gas be at 227°C and 600 torr of pressure?

1. _____

2. _____

3. _____

4. _____

5. _____

6. _____

NAME _____

THE GAS LAWS
EVALUATION TEST 4
PART 6

1. What is the pressure at STP?

2. What is the temperature at STP?

3. What is the molar volume of any gas at STP?

4. What are the numerical value and units of the gas constant?

5. If the number of moles of a gas are doubled at the same temperature and pressure, will the volume increase or decrease?

6. What volume will 1.27 moles of helium gas occupy at STP?

1. _____

2. _____

3. _____

4. _____

5. _____

6. _____

7. _____

8. _____

9. _____

7. At what pressure would 0.150 mole of nitrogen gas at 23°C occupy 8.90 L?

8. What volume would 32.0 g of NO_2 gas occupy at 3.12 atm and 18°C?

9. What is the density of 2.24 moles of NH_3 gas at 5°C and 3.22 atm?

NAME _____

THE GAS LAWS
EVALUATION TEST 5
PART 7–PART 8

1. How many liters of H_2 gas at STP will be produced by 0.400 mole of Ca in the following reaction?

$$3Ca + 2H_3PO_4 \longrightarrow Ca_3(PO_4)_2 + 3H_2$$

1. _____

2. _____

3. _____

4. _____

5. _____

2. How many liters of O_2 gas at STP will be produced by 54.0 g of HgO in the following reaction?

$$2HgO \longrightarrow 2Hg + O_2$$

3. What volume of HCl gas will be produced at 25.0°C and 2.75 atm from 5.0 g of H_2 in the following reaction?

$$H_2 + Cl_2 \longrightarrow 2HCl$$

4. What volume of HCl gas can be produced from 8.5 liters of H_2 in the following reaction?

$$H_2 + Cl_2 \longrightarrow 2HCl$$

5. 225 mL of N_2O gas are collected over water at 22°C and 790 torr. The amount of gas weighs 0.425 g and the vapor pressure of water at 22°C is 20 torr. What is the molecular weight of N_2O?

TEST ANSWERS

UNIT 1 *Whole Numbers and Decimals, Test 1*

1. 7 **2.** 9 **3.** 3 **4.** 0 **5.** 6 **6.** 4 **7.** 2 **8.** 1 **9.** 432 **10.** 701 **11.** 8300
12. 76,050 **13.** 214,408 **14.** 5,200,000 **15.** 18,000,612 **16.** 632,914,609

UNIT 1 *Whole Numbers and Decimals, Test 2*

1. 0 **2.** 0 **3.** four tenths **4.** seventeen hundredths **5.** three thousandths **6.** one hundred forty-two
thousandths **7.** 0.9 **8.** 0.04 **9.** 0.088 **10.** 0.0009 **11.** eight point two **12.** six point zero three
13. sixty-eight point zero four **14.** three point zero zero one two **15.** 3.6 **16.** 32.0 **17.** 72.03
18. 12.3

UNIT 1 *Whole Numbers and Decimals, Test 3*

1. 0.559 **2.** 47.2266 **3.** 5.3012 **4.** 6.46 **5.** 3.913 **6.** 0.081 **7.** 0.0328 **8.** 1.76076
9. 0.8 **10.** 2 **11.** 0.9 **12.** 0.41

UNIT 1 *Whole Numbers and Decimals, Test 4*

1. 51.06 **2.** 34.09 **3.** 35.3 **4.** 0.467 **5.** 82.899 **6.** 1.0266 **7.** 2.5 **8.** 4.5 **9.** 33.1
10. 1.898 **11.** 11.11 **12.** 0.417 **13.** 3.2 **14.** 3 **15.** 6.8 **16.** 3 **17.** 97 **18.** 200
19. 3.48 **20.** 5.2

UNIT 2 *Signed Numbers*

1. -9 **2.** -1 **3.** $+1$ **4.** $+2$ **5.** -2 **6.** -5 **7.** -8 **8.** -34 **9.** $+44$ **10.** $+30$
11. $+3$ **12.** -11 **13.** -7 **14.** $+9$ **15.** -4 **16.** -2 **17.** $+10$ **18.** $+5$ **19.** -3
20. $+6$ **21.** -6.4 **22.** $+6.88$

UNIT 3 *Powers of Ten*

1. 1×10^4 **2.** 1×10^0 **3.** 8.06×10^5 **4.** 2.19×10^5 **5.** 2.8×10^7 **6.** 1×10^{-4}
7. 1×10^{-1} **8.** 3.8×10^{-4} **9.** 1.01×10^{-2} **10.** 6.4×10^{-7} **11.** 2×10^7 **12.** 1.6×10^4
13. 8×10^{-5} **14.** 1.8×10^4 **15.** 8×10^{-2} **16.** 2.5×10^{-2} **17.** 4×10^2 **18.** 3×10^{-2}
19. 8×10^{-4} **20.** 2×10^4 **21.** 0.442 **22.** 3.75 **23.** -3.15 **24.** -2.05

UNIT 4 *Measurement Concepts and Dimensional Analysis*

1. 152.5 **2.** 1.735 **3.** 15.7 **4.** two **5.** four **6.** six **7.** 790 **8.** 2.13
9. 1.00 **10.** 3.1 **11.** 3 cm **12.** 60 L **13.** 4 cm **14.** 0.1 L **15.** 4000 g

UNIT 5 *The Metric System, Density and Specific Gravity, Test 1*

1. 1000 **2.** 100 **3.** 10 **4.** 10 **5.** 1000 **6.** 100 **7.** 1000 **8.** 100 **9.** 1000 **10.** 10
11. 10 m $\times \dfrac{100 \text{ cm}}{1 \text{ m}}$ **12.** 89 m $\times \dfrac{1 \text{ km}}{1000 \text{ m}}$ **13.** 19 cm $\times \dfrac{10 \text{ mm}}{1 \text{ cm}}$ **14.** 1860 mm **15.** 1.92 m
16. 0.150 m **17.** 51 cm **18.** 1.3 cm

UNIT 5   The Metric System, Density and Specific Gravity, Test 2

1. 1000   2. 10   3. 100   4. 1000   5. 100   6. 1000   7. 0.798 L 8. 3600 mL   9. 92 cm^3
10. 4400 g   11. 0.429 kg   12. 5600 mg   13. quantity of matter   14. gravity   15. balance
16. 10 mL   17. 28 g   18. 2.00 lb   19. 394 in   20. 3.0 kg   21. 4.75 L

UNIT 5   The Metric System, Density and Specific Gravity, Test 3

1. 212   2. 0   3. 100   4. 32   5. a degree Celsius   6. $C = \dfrac{°F - 32}{1.8}$ 7. $F = (1.8 \times °C) + 32$

8. $F = (1.8 \times °C) + 32$ 9. $C = \dfrac{°F - 32}{1.8}$ 10. $F = (1.8 \times °C) + 32$ 11. $C = \dfrac{°F - 32}{1.8}$
$\quad C = 30°C$ $\quad\quad F = 41°F$ $\quad C = -10°C$ $\quad\quad F = 16°F$
$\quad F = (1.8 \times 30°) + 32$ $\quad C = \dfrac{41° - 32}{1.8}$ $\quad F = (1.8 \times -10°) + 32$ $\quad C = \dfrac{16° - 32}{1.8}$
$\quad F = 54 + 32 = 86°F$ $\quad C = \dfrac{9}{1.8} = 5°C$ $\quad F = -18 + 32 = 14°F$ $\quad C = \dfrac{-16}{1.8} = -8.9°C$

UNIT 5   The Metric System, Density and Specific Gravity, Test 4

1. $d = \dfrac{420\,g}{250\ cm^3} = 1.68\ g/cm^3$ 2. $d = \dfrac{28.6\,g}{30.5\ mL} = 0.938\ g/mL$ 3. $11.6\ cm^3 \times \dfrac{19.3\,g}{cm^3} = 224\ g$

4. $20.40\ mL \times \dfrac{13.55\,g}{mL} = 276.4\ g$ 5. $\dfrac{115\,g}{150\ mL} = 0.767\ g/mL$ 6. $27.0\ cm^3 \times \dfrac{7.78\,g}{cm^3} = 210\ g$

UNIT 6   The Divisions and Properties of Matter

1. matter   2. element   3. atom   4. molecule   5. compound   6. solid   7. gas   8. physical
change   9. chemical change   10. molecule (or compound)   11. physical   12. chemical   13. B
14. Cl   15. Au   16. Hg   17. Zn   18. Ca   19. S   20. Cr   21. I   22. magnesium
23. copper   24. potassium   25. silver   26. tin   27. phosphorus   28. helium   29. lead
30. carbon

UNIT 7   Atomic Structure, Test 1

1. a. e^{-1} 1. b. -1 1. c. n 1. d. 0   1. e. proton   1. f. $+1$ 2. a. 20   2. b. 53
3. a. P (phosphorus)   3. b. Au (gold)   4. a. 12   4. b. 12   4. c. 12   4. d. Ca   4. e. 20
4. f. 20   4. g. Si   4. h. 14   4. i. 14   4. j. O   4. k. 8   4. l. 8   5. a. 8   5. b. 32

UNIT 7   Atomic Structure, Test 2

1. a. 3   1. b. 2 1   1. c. 7   1. d. 2 5   1. e. 12   1. f. 2 8 2   1. g. 16   1. h. 2 8 6
1. i. 20   1. j. 2 8 8 2   1. k. 35   1. l. 2 8 18 7   1. m. 33   1. n. 2 8 18 5   2. a. 2 8 1
2. b. Na·   2. c. 2 8 3   2. d. :A̤l   2. e. 2 6   2. f. ·Ö: 2. g. 2 8 7   2. h. :C̈l: 2. i. 2 8
2. j. :N̈e: 3. a. 2 1   3. b. one lost   3. c. Li^{1+} 3. d. 2 8 2   3. e. two lost   3. f. Mg^{2+}
3. g. 2 8 3   3. h. three lost   3. i. Al^{3+} 3. j. 2 5   3. k. three gained   3. l. N^{3-} 3. m. 2 6
3. n. two gained   3. o. O^{2-} 3. p. 2 7   3. q. one gained   3. r. F^{1-}

UNIT 7   Atomic Structure, Test 3

1. a. 2 8 8 1   1. b. IA   1. c. 2 5   1. d. VA   1. e. 2 8 8 2   1. f. IIA   1. g. 2 8 6
1. h. VIA   1. i. 2 8 7   1. j. VIIA   2. a. IA   2. b. Na·   2. c. VIA   2. d. ·Ö: 2. e. IIA
2. f. Mg: 2. g. VIIA   2. h. :B̈r: 2. i. IIIA   2. j. :A̤l   2. k. VA   2. l. :N̈·   3. a. VA
3. b. P^{3-} 3. c. IA   3. d. Na^{1+} 3. e. IIA   3. f. Mg^{2+} 3. g. VIA   3. h. S^{2-} 3. i. VIIA
3. j. F^{-1} 3. k. IIIA   3. l. B^{3+} 4. a. $1s^2 2s^2 2p^6 3s^2$ 4. b. $1s^2 2s^2 2p^6 3s^2 3p^4$ 4. c. $1s^2 2s^2 2p^6 3s^2 3p^6 4s^2 3d^3$
4. d. $1s^2 2s^2 2p^6 3s^2 3p^6 4s^2 3d^{10} 4p^6 5s^2$ 4. e. $1s^2 2s^2 2p^6 3s^2 3p^6 4s^2 3d^{10} 4p^6 5s^2 4d^{10} 5p^5$
4. f. $1s^2 2s^2 2p^6 3s^2 3p^6 4s^2 3d^{10} 4p^6 5s^2 4d^{10} 5p^6 6s^1$

UNIT 8   Chemical Bonding and Formula Writing, Test 1

1. transfer   2. covalent   3. a. Zn^{2+} 3. b. Pb^{2+} 3. c. Ag^{1+} 3. d. Mg^{2+} 3. e. K^{1+} 3. f. Al^{3+}
3. g. Ba^{2+} 3. h. Ca^{2+} 3. i. H^{1+} 3. j. Na^{1+} 4. a. O^{2-} 4. b. S^{2-} 4. c. F^{1-} 4. d. Cl^{1-}
4. e. N^{3-} 4. f. Br^{1-} 4. g. I^{1-} 4. h. P^{3-} 5. a. PO_4^{3-} 5. b. NH_4^{1+} 5. c. OH^{1-} 5. d. CO_3^{2-}
5. e. $C_2H_3O_2^{1-}$ 5. f. SO_4^{2-} 5. g. NO_3^{1-}

UNIT 8   Chemical Bonding and Formula Writing, Test 2

1. HCl 2. CaI_2 3. BaO 4. $AlBr_3$ 5. K_2O 6. Na_2S 7. AlN 8. Zn_3N_2 9. Al_2S_3
10. H_2SO_4 11. $Zn(NO_3)_2$ 12. $Al(OH)_3$ 13. $Ca(OH)_2$ 14. $Ba_3(PO_4)_2$ 15. $Mg(C_2H_3O_2)_2$
16. NH_4Cl 17. $Al_2(CO_3)_3$ 18. NH_4NO_3 19. $(NH_4)_2S$ 20. $(NH_4)_3PO_4$ 21. $(NH_4)_2SO_4$
22. $:\overset{..}{F}:\overset{..}{F}:$ 23. $H:\overset{..}{Br}:$

UNIT 9   Naming Compounds, Test 1

1. base   2. base   3. salt   4. acid   5. oxide   6. acid   7. a. common   7. b. water
8. a. simple   8. b. aluminum nitride   9. a. simple   9. b. sodium phosphide   10. a. variable
10. b. iron (III) oxide   11. a. variable   11. b. copper (I) chloride   12. a. variable   12. b. tin (IV) iodide
13. a. acid   13. b. hydrofluoric acid   14. a. acid   14. b. hydrosulfuric acid   15. a. acid
15. b. hydrobromic acid   16. a. nonmetal   16. b. nitrogen dioxide   17. a. nonmetal   17. b. diphosphorus
pentoxide   18. a. nonmetal   18. b. carbon dioxide   19. a. nonmetal   19. b. carbon tetrachloride
20. a. common   20. b. ammonia   21. a. nonmetal   21. b. carbon monoxide

UNIT 9   Naming Compounds, Test 2

1. a. binary   1. b. hydroiodic acid   2. a. ternary   2. b. sulfuric acid   3. a. binary   3. b. hydrosulfuric
acid   4. a. binary   4. b. hydrobromic acid   5. a. ternary   5. b. nitric acid   6. a. binary
6. b. hydrochloric acid   7. a. ternary   7. b. acetic acid   8. a. ternary   8. b. phosphoric acid
9. a. positive monatomic ion and polyatomic ion   9. b. calcium carbonate   10. a. polyatomic ion and negative
monatomic ion   10. b. ammonium sulfide   11. a. two polyatomic ions   11. b. ammonium phosphate
12. a. positive monatomic ion and polyatomic ion   12. b. copper (II) sulfate   13. a. positive monatomic ion and
polyatomic ion   13. b. iron (III) hydroxide   14. a. polyatomic ion and negative monatomic ion
14. b. ammonium chloride   15. a. two polyatomic ions   15. b. ammonium acetate   16. a. positive monatomic
ion and polyatomic ion   16. b. tin (II) nitrate   17. a. two polyatomic ions   17. b. ammonium hydroxide
18. a. two polyatomic ions   18. b. ammonium sulfate

UNIT 9   Naming Compounds, Test 3

1. salt with more than one positive ion   2. ternary oxy-salt   3. ternary oxy-acid   4. ternary oxy-acid
5. ternary oxy-salt   6. ternary oxy-acid   7. sulfuric acid   8. sulfurous acid   9. nitrous acid
10. nitric acid   11. chlorous acid   12. hypochlorous acid   13. perchloric acid   14. chloric acid
15. sodium bromate   16. sodium bromite   17. sodium hypobromite   18. sodium perbromate
19. potassium hydrogen carbonate   20. sodium aluminum sulfate   21. potassium hydrogen sulfide
22. calcium ammonium phosphate

UNIT 10   Balancing Chemical Equations, Test 1

1. a. 8   1. b. 12   1. c. 3   1. d. 6   1. e. 6   1. f. 4   1. g. 6   1. h. 24   2. H_2, O_2, N_2, F_2,
Cl_2, Br_2, I_2 3. a. $H_2 + Cl_2 \longrightarrow 2HCl$ 3. b. $2HgO \longrightarrow 2Hg + O_2$ 3. c. $2KClO_3 \longrightarrow 2KCl + 3O_2$
3. d. $4Na + CO_2 \longrightarrow C + 2Na_2O$ 3. e $2NaCl + Pb(NO_3)_2 \longrightarrow 2NaNO_3 + PbCl_2$ 4. a. lost   4. b. 4
4. c. gained   4. d. 1   4. e. lost   4. f. 8   5. a. loss   5. b. gain

UNIT 10   Balancing Chemical Equations, Test 2

1. Na and H_2 2. a. 2+   2. b. 2−   2. c. 1+   3. a. $Mg^{2+}Cl_2^{1-}$ 3. b. $Na_2^{1+}S^{2-}$ 3. c. $Fe^{2+}O^{2-}$
4. $H_2^0 + O_2^0 \longrightarrow H_2^{1+}O^{2-}$ 5. a. $H^{1+}N^{5+}O_3^{2-}$ 5. b. $Na_2^{1+}S^{6+}O_4^{2-}$ 5. c. $K^{1+}Cl^{7+}O_4^{2-}$
6. a. $H_2^{1+}S^{2-} + H^{1+}N^{5+}O_3^{2-} \longrightarrow S^0 + N^{2+}O^{2-} + H_2^{1+}O^{2-}$
6. b. $K^{1+}I^{1-} + Na^{1+}Cl^{1+}O^{2-} + H_2^{1+}O^{2-} \longrightarrow K^{1+}O^{2-}H^{1+} + Na^{1+}Cl^{1-} + I_2^0$
6. c. $K^{1+}I^{7+}O_4^{2-} + K^{1+}I^{1-} + H^{1+}Cl^{1-} \longrightarrow K^{1+}Cl^{1-} + I_2^0 + H_2^{1+}O^{2-}$

UNIT 10   Balancing Chemical Equations, Test 3

1. $P^0 + H^{1+}N^{5+}O_3^{2-} + H_2^{1+}O^{2-} \longrightarrow N^{2+}O^{2-} + H_3^{1+}P^{5+}O_4^{2-}$ **2. a.** gained   **2. b.** 1   **2. c.** lost   **2. d.** 2

3. a. $S^0 + 2H^{1+}N^{5+}O_3^{2-} \longrightarrow H_2^{1+}S^{6+}O_4^{2-} + 2N^{2+}O^{2-}$

    1 (−6e⁻)
    2 (+3e⁻)

3. b. $3H_2^{1+}S^{2-} + 2H^{1+}N^{5+}O_3^{2-} \longrightarrow 3S^0 + 2N^{2+}O^{2-} + 4H_2^{1+}O^{2-}$

    3 (−2e⁻)
    2 (+3e⁻)

3. c. $2H^{1+}N^{5+}O_3^{2-} + 6H^{1+}I^{1-} \longrightarrow 2N^{2+}O^{2-} + 3I_2^0 + 4H_2^{1+}O^{2-}$

    2 (+3e⁻)
    6 (−1e⁻)

3. d. $Ca^{2+}S^{2-} + I_2^0 + 2H^{1+}Cl^{1-} \longrightarrow Ca^{2+}Cl_2^{1-} + 2H^{1+}I^{1-} + S^0$

    1 (−2e⁻)
    2 (+1e⁻)

3. e. $3Cu^{2+}O^{2-} + 2N^{3-}H_3^{1+} \longrightarrow N_2^0 + 3H_2^{1+}O^{2-} + 3Cu^0$

    3 (+2e⁻)
    2 (−3e⁻)

UNIT 10   Balancing Chemical Equations, Test 4

1. $4Zn^0 + 10H^+ + 2N^{5+}O_3^{2-} \longrightarrow 4Zn^{2+} + N_2^{1+}O^{2-} + 5H_2^{1+}O^{2-}$

    4 (−2e⁻)
    2 (+4e⁻)

2. $2Mn^{7+}O_4^{2-} + 3S^{2-} + 4H_2^{1+}O^{2-} \longrightarrow 2Mn^{4+}O_2^{2-} + 3S^0 + 8O^{2-}H^{1+}$

    2 (+3e⁻)
    3 (−2e⁻)

3. $Ca^0 + 2H^+ \longrightarrow Ca^{2+} + H_2^0$

    1 (−2e⁻)
    2 (+1e⁻)

4. $2Al^0 + 6H^+ \longrightarrow 2Al^{3+} + 3H_2^0$

    2 (−3e⁻)
    6 (+1e⁻)

5. $2N^{5+}O_3^{2-} + 6Br^- + 8H^+ \longrightarrow 2N^{2+}O^{2-} + 3Br_2^0 + 4H_2O$

    2 (+3e⁻)
    6 (−1e⁻)

6. Decomposition

7. Combination

8. Single replacement

9. Double replacement

10. Decomposition

UNIT 11   Weight Relations in Chemistry, Test 1

1. 22   **2. a.** number   **2. b.** atomic masses   **3. a.** 24.3   **3. b.** 63.5   **3. c.** 32.1   **4. a.** Write the formula.   **4. b.** List the data.   **4. c.** Substitute the data into the formula.   **4. d.** Do the calculation.

5. a. number of moles = $\dfrac{\text{given weight in grams}}{\text{weight of 1 mole}}$

  b. given weight in grams = 80.25 g
    weight of 1 mole = 32.1 g

  c. number of moles = $\dfrac{80.25 \text{ g}}{32.1 \text{ g}}$

  d. number of moles = 2.50

6.

Type of Atom	Atomic Weight	Number of Atoms	Molecular Weight
N	14.0	3	42.0
H	1.0	12	12.0
P	31.0	1	31.0
O	16.0	4	64.0
			149.0

UNIT 11   *Weight Relations in Chemistry, Test 2*

1. $110 \text{ g MgO} \times \dfrac{1 \text{ mole}}{40.3 \text{ g}} = 2.73 \text{ moles}$ 2. atomic wt of N = 14.0   molecular wt of N_2 = 28.0

3. $30.1 \times 10^{23} \text{ atoms} \times \dfrac{1 \text{ mole}}{6.02 \times 10^{23} \text{ atoms}} = 5.00 \text{ moles}$

4. $1.17 \text{ moles} \times \dfrac{6.02 \times 10^{23} \text{ molecules}}{1 \text{ mole}} = 7.04 \times 10^{23} \text{ molecules}$

5.   Na: $\dfrac{23.0}{106.5} \times 100 = 21.6\%$

   Cl: $\dfrac{35.5}{106.5} \times 100 = 33.3\%$

   O: $\dfrac{48.0}{106.5} \times 100 = 45.1\%$

6.   Al: $\dfrac{22.1 \text{ g}}{27.0 \text{ g/mole}} = 0.819 \text{ mole}, \dfrac{0.819}{0.819} = 1.00$ The formula is $AlPO_4$

   P: $\dfrac{25.4 \text{ g}}{31.0 \text{ g/mole}} = 0.819 \text{ mole}, \dfrac{0.819}{0.819} = 1.00$

   O: $\dfrac{52.5 \text{ g}}{16.0 \text{ g/mole}} = 3.28 \text{ moles}, \dfrac{3.28}{0.819} = 4.00$

7.   C: $\dfrac{85.7 \text{ g}}{12.0 \text{ g/mole}} = 7.14 \text{ moles}, \dfrac{7.14}{7.14} = 1.00$ The empirical formula is CH_2

   H: $\dfrac{14.3 \text{ g}}{1.0 \text{ g/mole}} = 14.3 \text{ moles}, \dfrac{14.3}{7.14} = 2.00$

   $n = \dfrac{84.0 \text{ g/mole}}{14.0 \text{ g/mole}} = 6, (CH_2)_6 = C_6H_{12}$

8. $0.950 \text{ g} - 0.519 \text{ g} = 0.431 \text{ g}$ % of water $= \dfrac{0.431 \text{ g}}{0.950 \text{ g}} \times 100 = 45.4\%$

UNIT 12   *Chemical Equations and Stoichiometry, Test 1*

1. $2C_4H_{10}$ = 2 moles, $13O_2$ = 13 moles, $8CO_2$ = 8 moles, $10H_2O$ = 10 moles   2. a. Write the balanced equation.
2. b. Calculate the mole ratio.   2. c. Multiply the mole ratio by the number of moles of starting substance stated in
the problem.   3. $12.0 \text{ moles KClO}_3 \times \dfrac{3 \text{ moles O}_2}{2 \text{ moles KClO}_3} = 18.0 \text{ moles O}_2$ 4. number of moles $= \dfrac{\text{weight in grams}}{\text{weight of 1 mole}}$

5. weight in grams = (weight of 1 mole)(number of moles)   6. $2.00 \text{ moles Ca(OH)}_2 \times \dfrac{74.1 \text{ g}}{1 \text{ mole Ca(OH)}_2} = 148 \text{ g}$

7. $12.0 \text{ moles NaClO}_3 \times \dfrac{3 \text{ moles O}_2}{2 \text{ moles NaClO}_3} \times \dfrac{32.0 \text{ g}}{1 \text{ mole O}_2} = 576 \text{ g}$

UNIT 12   *Chemical Equations and Stoichiometry, Test 2*

1. $3.5 \text{ moles AgNO}_3 \times \dfrac{1 \text{ mole Cu}}{2 \text{ moles AgNO}_3} = 1.75 \text{ moles Cu}$ 2. a. Convert grams of starting substance to moles of
starting substance.   2. b. Calculate the mole ratio.   2. c. Multiply the moles of starting substance by the mole
ratio.   2. d. Change moles of desired substance to grams.

3. $120 \text{ g Na}_2O \times \dfrac{1 \text{ mole Na}_2O}{62.0 \text{ g Na}_2O} \times \dfrac{2 \text{ moles NaOH}}{1 \text{ mole Na}_2O} \times \dfrac{40.0 \text{ g NaOH}}{1 \text{ mole NaOH}} = 155 \text{ g NaOH}$

4. moles of K $= \dfrac{250 \text{ g}}{39.1 \text{ g/mole}} = 6.39$ moles

moles of $Cl_2 = \dfrac{100 \text{ g}}{71.0 \text{ g/mole}} = 1.41$ moles

ratio of moles calculated $= \dfrac{6.39/1.41}{1.41/1.41} = \dfrac{4.53 \text{ moles K}}{1.00 \text{ mole } Cl_2}$

mole ratio from equation $= \dfrac{2 \text{ moles K}}{1 \text{ mole } Cl_2}$

K is in excess, so Cl_2 is the limiting reagent.

1.41 moles $Cl_2 \times \dfrac{2 \text{ moles KCl}}{1 \text{ mole } Cl_2} \times \dfrac{74.6 \text{ g KCl}}{1 \text{ mole KCl}} = 210$ g KCl

UNIT 13   The Concentration of Solutions, Test 1

1. a   2. a   3. b   4. a   5. b   6. b   7. c   8. d

9. percentage by weight $= \dfrac{\text{grams of solute}}{\text{grams of solution}} \times 100$ 10. b and c   11. grams of solute + grams of solvent

12. a. 30 g   12. b. 50 g   12. c. 80 g

UNIT 13   The Concentration of Solutions, Test 2

1. b   2. a

3. a. percentage by weight $= \dfrac{\text{g solute}}{\text{g solution}} \times 100$

b. g solute = 5 g
   g solution = 40 g

c. percentage by weight $= \dfrac{5.0 \text{ g}}{40 \text{ g}} \times 100$

d. percentage by weight = 13%

4. a. percentage by weight $= \dfrac{\text{g solute}}{\text{g solution}} \times 100$

b. g solute = 20 g
   g solution = 20 g + 80 g = 100 g

c. percentage by weight $= \dfrac{20 \text{ g}}{100 \text{ g}} \times 100$

d. percentage by weight = 20%

5. a. percentage by volume $= \dfrac{\text{volume of solute}}{\text{volume of solution}} \times 100$

b. volume of solute = 20 mL
   volume of solution = 200 mL

c. percentage by volume $= \dfrac{20 \text{ mL}}{200 \text{ mL}} \times 100$

d. percentage by volume = 10%

6. a. $M = \dfrac{\text{moles of solute}}{\text{liters of solution}}$

b. moles = 4
   liters = 5

c. $M = \dfrac{4 \text{ moles}}{5 \text{ } L}$

d. $M = 0.8$ moles/L

UNIT 13   The Concentration of Solutions, Test 3

1. c   2. b

3. a. number of moles $= \dfrac{\text{given weight in grams}}{\text{weight of 1 mole}}$

b. weight in grams = 0.5
   weight of 1 mole of Ca = 40.1 $\times$ 1 = 40.1 g
   weight of 1 mole of C = 12.0 $\times$ 1 = 12.0 g
   weight of 1 mole of O = 16.0 $\times$ 3 = $\underline{48.0 \text{ g}}$
                                    100.1 g

c. number of moles $= \dfrac{0.5}{100.1}$

d. number of moles = 0.005

aa. $M = \dfrac{\text{moles of solute}}{\text{liters of solution}}$

bb. moles = 0.005
    liters = 2

cc. $M = \dfrac{0.005 \text{ mole}}{2 \text{ } L}$

dd. $M = 0.003$ mole/L

4. a. $C_1V_1 = C_2V_2$ b. $\dfrac{C_1V_1}{V_2} = \dfrac{C_2\cancel{V_2}}{\cancel{V_2}}$ c. $C_2 = \dfrac{C_1V_1}{V_2}$

5. a. $C_1V_1 = C_2V_2 \longrightarrow C_2 = \dfrac{C_1V_1}{V_2}$

    b. $C_1 = 0.8\,M$ c. $C_2 = \dfrac{(0.8\ M)(80\ mL)}{320\ mL}$

       $V_1 = 80$ mL

       $C_2 = ?$ d. $C_2 = 0.2\,M$

       $V_2 = 320$ mL

6. a. $C_1V_1 = C_2V_2 \longrightarrow V_2 = \dfrac{C_1V_1}{C_2}$

    b. $C_1 = 20\%$ c. $V_2 = \dfrac{(20\%)(20\ mL)}{5\%}$

       $V_1 = 20$ mL

       $C_2 = 5\%$ d. $V_2 = 80$ mL

       $V_2 = ?$ e. $V_2 - V_1$

                      $= 80$ mL $- 20$ mL $= 60$ mL

UNIT 13   The Concentration of Solutions, Test 4

1. $\dfrac{15.0\ g\ KBr}{175\ mL} \times \dfrac{1\ mole\ KBr}{119\ g\ KBr} \times \dfrac{1000\ mL}{1\ L} = 0.720$ mole/L

2. $\dfrac{0.280\ mole}{1\ L} \times \dfrac{600\ mL}{1000\ mL/L} = 0.168$ mole

3. 0.625 mole KOH $\times \dfrac{1\ L}{0.850\ mole\ KOH} \times \dfrac{1000\ mL}{1\ L} = 735$ mL

4. $\dfrac{0.550\ mole}{1\ L} \times \dfrac{50.0\ mL}{1000\ mL/L} \times \dfrac{85.0\ g}{1\ mole} = 2.34$ g

5. 10.5 g $\times \dfrac{1\ mole}{53.5\ g\ NH_4Cl} \times \dfrac{1\ L}{0.320\ mole} \times \dfrac{1000\ mL}{1\ L} = 613$ mL

UNIT 13   The Concentration of Solutions, Test 5

1. $\dfrac{98.1\ g\ H_2SO_4}{2} = 49.1$ g   2. 21.0 g HNO$_3$ $\times \dfrac{1\ equiv}{63.0\ g} = 0.333$ equiv

3. Number of equivalents $= \dfrac{5.60\ g}{32.7\ g/equiv} = 0.171$ equiv $N = \dfrac{0.171\ equiv}{0.100\ L} = 1.71$ equiv/L

4. $C_2 = \dfrac{(0.475\ M)(21.9\ mL)}{28.1\ mL)} = 0.370\,M$ 5. a. 6.17–acidic   b. 11.0–basic

UNIT 14   The Gas Laws, Test 1

1. a. 760   1. b. 1   2. 0.5   3. 1520   4. decrease   5. double   6. a. $P_1V_1 = P_2V_2$

6. b. $\dfrac{P_1V_1}{P_2} = \dfrac{P_2V_2}{P_2}$ 6. c. $\dfrac{P_1V_1}{P_2} = \dfrac{P_2V_2}{P_2}$ 6. d. $V_2 = \dfrac{P_1V_1}{P_2}$ 7. a. $P_1V_1 = P_2V_2 \longrightarrow V_2 = \dfrac{P_1V_1}{P_2}$

7. b. $P_1 = 8$ atm $V_1 = 600$ mL $P_2 = 2$ atm $V_2 = ?$ 7. c. $V_2 = \dfrac{(8\ atm)(600\ mL)}{2\ atm}$ 7. d. $V_2 = 2400$ mL

8. a. $P_1V_1 = P_2V_2 \longrightarrow V_2 = \dfrac{P_1V_1}{P_2}$ 8. b. $P_1 = 800$ torr $V_1 = 400$ mL $P_2 = 1000$ torr $V_2 = ?$

8. c. $V_2 = \dfrac{(800\ torr)(400\ mL)}{1000\ torr}$ 8. d. $V_2 = 320$ mL   9. a. $P_1V_1 = P_2V_2 \longrightarrow V_2 = \dfrac{P_1V_1}{P_2}$

9. b. $P_1 = 6$ atm $V_1 = 4\,L$ $P_2 = 2$ atm $V_2 = ?$ 9. c. $V_2 = \dfrac{(6\ atm)(4\ L)}{2\ atm}$ 9. d. $V_2 = 12\,L$

UNIT 14   The Gas Laws, Test 2

1. $373\,K$ 2. absolute zero   3. decrease   4. two   5. $40^\circ C + 273 = 313\,K$

6. $-20^\circ C + 273 = 253\,K$ 7. $\dfrac{V_1}{T_1} = \dfrac{V_2}{T_2} \longrightarrow \dfrac{V_1T_2}{T_1} = \dfrac{V_2\cancel{T_2}}{\cancel{T_2}} \longrightarrow V_2 = \dfrac{V_1T_2}{T_1}$

8. a. $\dfrac{V_1}{T_1} = \dfrac{V_2}{T_2} \longrightarrow V_2 = \dfrac{V_1 T_2}{T_1}$ **8. b.** $V_1 = 6.0\ L$ $T_1 = 300\,K$ $V_2 = ?$ $T_2 = 423\,K$

8. c. $V_2 = \dfrac{(6.0\ L)(423\,K)}{300\,K}$ **8. d.** $V_2 = 8.5\ L$ **9. a.** $\dfrac{V_1}{T_1} = \dfrac{V_2}{T_2} \longrightarrow V_2 = \dfrac{V_1 T_2}{T_1}$

9. b. $V_1 = 400\ mL$ $T_1 = 225°C + 273 = 498\,K$ $V_2 = ?$ $T_2 = 400\,K$ **9. c.** $V_2 = \dfrac{(400\ mL)(400\,K)}{498\,K}$

9. d. $V_2 = 321\ mL$ **10. a.** $\dfrac{V_1}{T_1} = \dfrac{V_2}{T_2} \longrightarrow V_2 = \dfrac{V_1 T_2}{T_1}$ **10. b.** $V_1 = 8.0\ L$ $T_1 = 210°C + 273 = 483\,K$

$V_2 = ?$ $T_2 = 250\,K$ **10. c.** $V_2 = \dfrac{(8.0\ L)(250\,K)}{483\,K}$ **10. d.** $V_2 = 4.1\ L$

UNIT 14   The Gas Laws, Test 3

1. b   **2.** c   **3.** $\dfrac{V_1}{T_1} = \dfrac{V_2}{T_2}$ **4.** $\dfrac{V_1 P_1}{T_1} = \dfrac{V_2 P_2}{T_2}$ **5. a.** $\dfrac{P_1 V_1}{T_1} = \dfrac{P_2 V_2}{T_2} \longrightarrow P_2 V_2 T_1 = P_1 V_1 T_2$

5. b. $\dfrac{P_2 V_2 T_1}{P_2 T_1} = \dfrac{P_1 V_1 T_2}{P_2 T_1}$ **5. c.** $\dfrac{P_2 V_2 T_1}{P_2 T_1} = \dfrac{P_1 V_1 T_2}{P_2 T_1}$ **5. d.** $V_2 = \dfrac{P_1 V_1 T_2}{P_2 T_1}$ **6. a.** $V_2 = \dfrac{P_1 V_1 T_2}{P_2 T_1}$

6. b. $P_1 = 300\ torr$ $V_1 = 800\ mL$ $T_1 = -23°C + 273 = 250\,K$ $P_2 = 600\ torr$ $V_2 = ?$ $T_2 = 227°C + 273 = 500\,K$

6. c. $V_2 = \dfrac{(300\ torr)(800\ mL)(500\,K)}{(600\ torr)(250\,K)}$ **6. d.** $V_2 = 800\ mL$

UNIT 14   The Gas Laws, Test 4

1. 760 torr, or 1 atm   **2.** 0°C, or 273 K   **3.** 22.4 L   **4.** 0.0821 L-atm/mole-K   **5.** increase

6. a. $PV = nRT \longrightarrow V = \dfrac{nRT}{P}$

b. $n = 1.27$ moles
$R = 0.0821 L$-atm/mole-K
$T = 273\,K$
$P = 1.00$ atm

c. $V = \dfrac{(1.27\ moles)(0.0821\ L\text{-}atm)(273\,K)}{(1.00\ atm)(mole\text{-}K)}$

d. $V = 28.5\ L$

7. a. $PV = nRT \longrightarrow P = \dfrac{nRT}{V}$

b. $n = 0.150$ mole
$R = 0.0821\ L$-atm/mole-K
$T = 23°C + 273 = 296\,K$
$V = 8.90\ L$

c. $P = \dfrac{(0.150\ mole)(0.0821\ L\text{-}atm)(296\,K)}{(8.90\ L)(mole\text{-}K)}$

d. $P = 0.410$ atm

8. a. $PV = nRT \longrightarrow V = \dfrac{nRT}{P}$

b. number of moles $= \dfrac{\text{given weight in grams}}{\text{weight of 1 mole}}$

weight in grams $= 32.0\ g$
weight of 1 mole of N $= 14.0 \times 1 = 14.0\ g$
weight of 1 mole of O $= 16.0 \times 2 = \dfrac{32.0\ g}{46.0\ g}$

number of moles $\dfrac{32.0\ g}{46.0\ g} = 0.696$ mole

$R = 0.0821\ L$-atm/mole-K
$T = 18°C + 273 = 291\,K$
$P = 3.12$ atm

c. $V = \dfrac{(0.696\ mole)(0.0821\ L\text{-}atm)(291\,K)}{(3.12\ atm)(mole\text{-}K)}$

d. $V = 5.33\ L$

9. a. $PV = nRT \longrightarrow V = \dfrac{nRT}{P}$

b. $n = 2.24$ moles
$R = 0.0821\ L$-atm/mole-K
$T = 5°C + 273 = 278\,K$
$P = 3.22$ atm

c. $V = \dfrac{(2.24\ moles)(0.0821\ L\text{-}atm)(278\,K)}{(3.22\ atm)(mole\text{-}K)}$

d. $V = 15.9\ L$

aa. density $= \dfrac{\text{mass in grams}}{\text{volume in liters}}$

bb. mass in grams $= $ (number of moles)(weight of 1 mole)
number of moles $= 2.24$
weight of 1 mole of N $= 14.0 \times 1 = 14.0\ g$
weight of 1 mole of H $= 1.0 \times 3 = \dfrac{3.0\ g}{17.0\ g}$

mass in grams $= $ (2.24 moles)(17.0 g/mole)
mass in grams $= 38.1\ g$

cc. $V = 15.9\ L$

dd. density $= \dfrac{38.1\ g}{15.9\ L} = 2.40\ g/L$

UNIT 14   The Gas Laws, Test 5

1. $0.400 \text{ mole Ca} \times \dfrac{3 \text{ moles H}_2}{3 \text{ moles Ca}} \times \dfrac{22.4 \; L \; \text{H}_2}{1 \text{ mole H}_2} = 8.96 \; L\text{H}_2$

2. $54.0 \text{ g HgO} \times \dfrac{1 \text{ mole HgO}}{216.6 \text{ g HgO}} \times \dfrac{1 \text{ mole O}_2}{2 \text{ moles HgO}} \times \dfrac{22.4 \; L}{1 \text{ mole O}_2} = 2.79 \; L$

3. $5.0 \text{ g H}_2 \times \dfrac{1 \text{ mole H}_2}{2.0 \text{ g H}_2} \times \dfrac{2 \text{ moles HCl}}{1 \text{ mole H}_2} = 5.0$ moles of HCl produced

$V = \dfrac{(5.0 \text{ moles H}_2)\,(0.0821 \; L\text{-atm})\,(298 \text{ K})}{(2.75 \text{ atm})(\text{mole-K})} = 44 \; L$ 4. $8.5 \; L\,\text{H}_2 \times \dfrac{2 \text{ volumes HCl}}{1 \text{ volume H}_2} = 17 \; L \; \text{HCl}$

5. $\text{MW} = \dfrac{(0.425 \text{ g})(0.0821 \; L\text{-atm})(295 \text{ K})}{(1.04 \text{ atm})(0.225 \; L\text{-atm})(\text{mole-K})} = 44.0$

Periodic Table of the Elements

Legend:
- Atomic Number: 11
- Name: Sodium
- Symbol: Na
- Atomic Mass: 23.0

Atomic masses are rounded off to the nearest 0.1.
[a] Mass number of most stable or best-known isotope.
[b] Mass of most commonly available long-lived isotope.

Period	IA	IIA	IIIB	IVB	VB	VIB	VIIB	VIII	VIII	VIII	IB	IIB	IIIA	IVA	VA	VIA	VIIA	Noble Gases
1	1 H 1.0																	2 He 4.0
2	3 Li 6.9	4 Be 9.0											5 B 10.8	6 C 12.0	7 N 14.0	8 O 16.0	9 F 19.0	10 Ne 20.2
3	11 Na 23.0	12 Mg 24.3											13 Al 27.0	14 Si 28.1	15 P 31.0	16 S 32.1	17 Cl 35.5	18 Ar 39.9
4	19 K 39.1	20 Ca 40.1	21 Sc 45.0	22 Ti 47.9	23 V 50.9	24 Cr 52.0	25 Mn 54.9	26 Fe 55.8	27 Co 58.9	28 Ni 58.7	29 Cu 63.5	30 Zn 65.4	31 Ga 69.7	32 Ge 72.6	33 As 74.9	34 Se 79.0	35 Br 79.9	36 Kr 83.8
5	37 Rb 85.5	38 Sr 87.6	39 Y 88.9	40 Zr 91.2	41 Nb 92.9	42 Mo 95.9	43 Tc 98.9b	44 Ru 101.0	45 Rh 102.9	46 Pd 106.4	47 Ag 107.9	48 Cd 112.4	49 In 114.8	50 Sn 118.7	51 Sb 121.8	52 Te 127.6	53 I 126.9	54 Xe 131.3
6	55 Cs 132.9	56 Ba 137.3	57 La 138.9 *	72 Hf 178.5	73 Ta 180.9	74 W 183.8	75 Re 186.2	76 Os 190.2	77 Ir 192.2	78 Pt 195.1	79 Au 197.0	80 Hg 200.6	81 Tl 204.4	82 Pb 207.2	83 Bi 208.9	84 Po (210)a	85 At (210)a	86 Rn (222)a
7	87 Fr (223)a	88 Ra 226.0	89 Ac (227)a **	104 Unq (261)a	105 Unp (262)a	106 Unh (263)a	107 Uns (262)a	108 Uno (265)a	109 Une (266)a									

Lanthanide Series * (Period 6)

58 Ce 140.1	59 Pr 140.9	60 Nd 144.2	61 Pm (145)a	62 Sm 150.4	63 Eu 152.0	64 Gd 157.2	65 Tb 158.9	66 Dy 162.5	67 Ho 164.9	68 Er 167.3	69 Tm 168.9	70 Yb 173.0	71 Lu 175.0

Actinide Series ** (Period 7)

90 Th 232.0b	91 Pa 231.0b	92 U 238.0	93 Np 237.0b	94 Pu (242)a	95 Am (243)a	96 Cm (247)a	97 Bk (249)a	98 Cf (251)a	99 Es (254)a	100 Fm (253)a	101 Md (256)a	102 No (254)a	103 Lr (257)a

GLOSSARY

absolute zero *See* Kelvin scale.

acid (1) A substance that produces H^+ (H_3O^+) when dissolved in water. (2) A proton donor. (3) An electron-pair acceptor. A substance that bonds to an electron pair.

atmospheric pressure The pressure experienced by objects on the earth as a result of the layer of air surrounding our planet. A pressure of 1 atmosphere (1 atm) is the pressure that will support a column of mercury 760 mm high at 0°C.

atom The smallest particle of an element that can enter into a chemical reaction.

atomic mass unit (amu) A unit of mass equal to one-twelfth the mass of a carbon-12 atom.

atomic number The number of protons in the nucleus of an atom.

atomic theory The theory that substances are composed of atoms, and that chemical reactions are explained by the properties and the interactions of these atoms.

atomic mass The average relative mass of the isotopes of an element referred to the atomic mass of carbon-12 as exactly 12 amu.

Avogadro's law Equal volumes of different gases at the same temperature and pressure contain equal numbers of molecules.

Avogadro's number 6.022×10^{23}; the number of formula units in 1 mole of whatever is indicated by the formula.

balanced equation A chemical equation having the same number and kind of atoms and the same electrical charge on each side of the equation.

barometer A device used to measure pressure.

base (1) A substance that produces OH^- when dissolved in water. (2) a proton acceptor. (3) An electron-pair donor.

binary compound A compound composed of two different elements.

boiling point The temperature at which the vapor pressure of a liquid is equal to the pressure above the liquid.

bond dissociation energy The energy required to break a covalent bond.

bond length The distance between two nuclei that are joined by a chemical bond.

Boyle's law At constant temperature, the volume of a given mass of gas is inversely proportional to the pressure (PV = constant).

buffer solution A solution that resists changes in pH when diluted or when small amounts of a strong acid or strong base are added.

calorie (cal) One calorie is a quantity of heat energy that will raise the temperature of 1 gram of water 1°C (from 14.5 to 15.5°C).

Celsius scale (°C) The temperature scale on which water freezes at 0°C and boils at 100°C at 1 atm pressure.

Charles' law At constant pressure, the volume of a gas is directly proportional to the absolute (K) temperature (V/T = constant).

chemical bond The attractive force that holds atoms together in a compound.

chemical change A change producing products that differ in composition from the original substances.

chemical equation An expression showing the reactants and the products of a chemical change (for example, $2H_2 + O_2 \rightarrow 2H_2O$).

chemical equilibrium The state in which the rate of the forward reaction equals the rate of the reverse reaction for a chemical change.

chemical family *See* groups or families of elements.

chemical formula A shorthand method for showing the composition of a compound using symbols of the elements.

chemical kinetics The study of reaction rates or the speed of a particular reaction.

chemical properties Properties of a substance related to its chemical changes.

chemistry The science dealing with the composition of matter and the changes in composition that matter undergoes.

colligative properties Properties of a solution that depend on the number of solute particles in solution and not on the nature of the solute (for example, vapor pressure lowering, freezing point lowering, boiling point elevation).

combination reaction A direct union or combination of two substances to produce one new substance.

combustion   In general, the process of burning or uniting a substance with oxygen, which is accompanied by the evolution of light and heat.

compound   A substance composed of two or more elements combined in a definite proportion by weight.

concentrated solution   A solution containing a relatively large amount of solute.

concentration of a solution   A quantitative expression of the amount of dissolved solute in a certain quantity of solvent or solution.

conjugate acid-base   Two molecules or ions whose formulas differ by one H^+. (The acid is the species with the H^+, and the base is the species without the H^+.)

coordinate-covalent bond   A covalent bond in which the shared pair of electrons is furnished by only one of the bonded atoms.

covalent bond   A chemical bond formed between two atoms by sharing a pair of electrons.

Dalton's atomic theory   The first modern atomic theory to state that elements are composed of tiny, individual particles called atoms.

decomposition reaction   A breaking down or decomposition of one substance into two or more different substances.

deliquescence   The absorption of water by a compound beyond the hydrate stage to form a solution.

density   The mass of an object divided by its volume.

diffusion   The process by which gases and liquids mix spontaneously because of the random motion of their particles.

dilute solution   A solution containing a relatively small amount of solute.

double bond   A covalent bond in which two pairs of electrons are shared.

double-displacement reaction   A reaction of two compounds to produce two different compounds by exchanging the components of the reacting compounds.

electron   A subatomic particle that exists outside the nucleus and has an assigned electrical charge of -1.

electron affinity   The energy released or absorbed when an electron is added to an atom or an ion.

electron-dot structure   *See* Lewis structure.

electronegativity   The relative attraction that an atom has for the electrons in a covalent bond.

electron shell   *See* energy levels of electrons.

element   A basic building block of matter that cannot be broken down into simpler substances by ordinary chemical changes.

empirical formula   A chemical formula that gives the smallest whole-number ratio of atoms in a compound.

energy levels of electrons   Areas in which electrons are located at various distances from the nucleus.

energy sublevels   The *s, p, d,* and *f* orbitals within a principal energy level occupied by electrons in an atom.

equivalent weight   That weight of a substance that will react with, combine with, contain, replace, or in any other way be equivalent to 1 mole of hydrogen atoms or hydrogen ions.

Fahrenheit scale (°F)   The temperature scale on which water freezes at 32°F and boils at 212°F at 1 atm pressure.

formula weight   The sum of the atomic weights of all the atoms in a chemical formula.

freezing or melting point   The temperature at which the solid and liquid states of a substance are in equilibrium.

gas   The state of matter that is the least compact of the three physical states; a gas has no shape or definite volume and completely fills its container.

groups or families of elements   Vertical groups of elements in the periodic table (IA, IIA, and so on). Families of elements have similar outer-orbital electron structures.

halogen family   Group VIIA of the periodic table; consists of the elements fluorine, chlorine, bromine, iodine, and astatine.

heat   A form of energy associated with the motion of small particles of matter.

heat of fusion   The amount of heat required to change 1 gram of a solid into a liquid at its melting point.

heat of reaction   The quantity of heat produced by a chemical reaction.

heat of vaporization   The amount of heat required to change 1 gram of a liquid to a vapor at its normal boiling point.

heterogeneous   Matter without uniform composition; two or more phases present.

homogeneous   Matter having uniform properties throughout.

hydrogen bond   A chemical bond between polar molecules that contain hydrogen covalently bonded to the highly electronegative atoms F, O, or N:

hydrolysis   A chemical reaction with water in which the water molecule is split into H^+ and OH^-

ideal gas   A gas that obeys the gas laws and the Kinetic-Molecular Theory exactly.

ideal gas equation $PV = nRT$; a single equation relating the four variables—*P, V, T,* and *n*—used in the gas laws. *R* is a proportionality constant known as the ideal or universal gas constant.

immiscible   Incapable of mixing. Immiscible liquids do not form solutions with one another.

inorganic chemistry   The chemistry of the elements and their compounds other than the carbon compounds.

ion   An electrically charged atom or group of atoms. A positively charged (+) ion is called a *cation*, and a negatively charged (−) ion is called an *anion*.

ionic bond A chemical bond between a positively charged ion and a negatively charged ion.

ionization The formation of ions.

isotopes Atoms of an element having the same atomic number but different atomic masses. Since the atomic numbers are identical, isotopes vary only in the number of neutrons in the nucleus.

IUPAC International Union of Pure and Applied Chemistry.

Kelvin (absolute) scale (K) Absolute temperature scale starting at absolute zero, the lowest temperature possible. Freezing and boiling points of water on this scale are 273 K and 373 K, respectively, at 1 atm pressure.

kilocalorie (kcal) 1000 cal; the kilocalorie is also known as the nutritional or large Calorie, used for measuring the energy produced by food.

kilogram (kg) The standard unit of mass in the metric system.

kinetic energy (KE) Energy of motion: $KE = \frac{1}{2}mv^2$.

Kinetic-Molecular Theory A group of assumptions used to explain the behavior and properties of ideal gas molecules.

law A statement of the occurrence of natural phenomena that occur with unvarying uniformity under the same conditions.

Law of Conservation of Energy Energy cannot be created or destroyed, but it may be transformed from one form to another.

Law of Conservation of Mass There is no detectable change in the total mass of the substances in a chemical reaction; the mass of the products equals the mass of the reactants.

Law of Definite Composition A compound always contains the same elements in a definite proportion by weight.

Le Chatelier's principle If the conditions of an equilibrium system are altered, the system will shift to establish a new equilibrium system under the new set of conditions.

Lewis structure A method of indicating the covalent bonds between atoms in a molecule or an ion such that a pair of electrons (:) represents the valence electrons forming the covalent bond.

limiting reactant A reactant in a chemical reaction that limits the amount of product formed. The limitation is imposed because an insufficient quantity of the reactant, compared to amounts of the other reactants, was used in the reaction.

liquid One of the three physical states of matter. The particles in a liquid move about freely while the liquid still retains a definite volume. Thus, liquids flow and take the shape of their containers.

liter (L) A unit of volume commonly used in chemistry; 1 L = 1000 ml; the volume of a kilogram of water at 4°C.

mass The quantity or amount of matter that an object possesses.

matter Anything that has mass and occupies space.

metal An element that is lustrous, ductile, malleable, and a good conductor of heat and electricity. Metals tend to lose their valence electrons and become positive ions.

metalloid An element having properties that are intermediate between those of metals and nonmetals.

meter (m) The standard unit of length in the SI and metric systems.

metric system A decimal system of measurements.

miscible Capable of mixing and forming a solution.

mixture Matter containing two or more substances that can be present in variable amounts.

molality (m) The number of moles of solute dissolved in 1000 grams of solvent.

molarity (M) The number of moles of solute per liter of solution.

molar solution A solution containing 1 mole of solute per liter of solution.

molar volume of a gas The volume of 1 mole of a gas at STP, 22.4 L/mol.

mole The amount of a substance containing the same number of formula units (6.022×10^{23}) as there are in exactly 12 grams of carbon-12. One mole is equal to the formula weight in grams of any substance.

molecular formula The true formula representing the total number of atoms of each element present in one molecule of a compound.

molecular mass The sum of the atomic masses of all the atoms in a molecule.

molecule A small, uncharged individual unit of a compound formed by the union of two or more atoms.

mole ratio A ratio of the number of moles of any two species in a balanced chemical equation. The mole ratio can be used as a conversion factor in stoichiometric calculations.

monomer The small unit or units that undergo polymerization to form a polymer.

monosaccharide A carbohydrate that cannot be hydrolyzed to simpler carbohydrate units: for example, simple sugars like glucose or fructose.

net ionic equation A chemical equation that includes only those molecules and ions that have changed in the chemical reaction.

neutralization The reaction of an acid and a base to form water plus a salt.

neutron A subatomic particle that is electrically neutral and has an assigned mass of 1 amu.

noble gases A family of elements in the periodic table—helium, neon, argon, krypton, and xenon—that contain a particularly stable electron structure.

nonelectrolyte A substance whose aqueous solutions do not conduct electricity.

nonmetal   Any of a number of elements that do not have the characteristics of metals. They are located mainly in the upper right-hand corner of the periodic table.

nonpolar covalent bond   A covalent bond between two atoms with the same electronegativity value. Thus, the electrons are shared equally between the two atoms.

normal boiling point   The temperature at which the vapor pressure of a liquid equals 1 atm or 760 torr pressure.

normality   The number of equivalent weights (equivalents) of solute per liter of solution.

nucleus   The central part of an atom where all the protons and neutrons of the atom are located. The nucleus is very dense and has a positive electrical charge.

octet rule   An atom tends to lose or gain electrons until it has eight electrons in its outer shell.

orbital   A cloudlike region around the nucleus where electrons are located. Orbitals are considered to be energy sublevels within the principal levels and are labeled *s, p, d,* and *f.*

oxidation   An increase in the oxidation number of an atom as a result of losing electrons.

oxidation number (oxidation state)   A small number representing the state of oxidation of an atom. For an ion, it is the positive or negative charge on the ion; for covalently bonded atoms, it is a positive or negative number assigned to the more electronegative atom; in free elements, it is zero.

oxidation-reduction   A chemical reaction wherein electrons are transferred from one element to another.

oxidizing agent   A substance that causes an increase in the oxidation state of another substance. The oxidizing agent is reduced during the course of the reaction.

percentage composition of a compound   The weight percent of each element in a compound.

percent yield $\dfrac{\text{Actual yield}}{\text{Theoretical yield}} \times 100\%$

periodic law   The properties of the chemical elements are a periodic function of their atomic numbers.

periodic table   An arrangement of the elements according to their atomic numbers, illustrating the periodic law. The table consists of horizontal rows or periods and vertical columns or families of elements. Each period ends with a noble gas.

periods of elements   The horizontal groupings of elements in the periodic table.

pH   A method of expressing the H^+ concentration (acidity) of a solution. $pH = -\log [H^+]$; $pH = 7$ is a neutral solution, $pH < 7$ is acidic, and $pH > 7$ is basic.

phase   A homogeneous part of a system separated from other parts by a physical boundary.

physical change   A change in form (such as size, shape, physical state) without a change in composition.

physical properties   Characteristics associated with the existence of a particular substance. Inherent characteristics such as color, taste, density, and melting point are physical properties of various substances.

physical states of matter   Solids, liquids, and gases.

pOH   A method of expressing the basicity of a solution. $pOH = -\log [OH^-]$. $pOH = 7$ is a neutral solution, $pOH < 7$ is basic, and $pOH > 7$ is acidic.

polar covalent bond   A covalent bond between two atoms with differing electronegativity values resulting in unequal sharing of bonding electrons.

polyatomic ion   An ion composed of more than one atom.

pressure   Force per unit area; expressed in many units, such as mm Hg, atm, in./cm², torr.

product   A chemical substance produced from reactants by a chemical change.

properties   The characteristics, or traits, of substances. Properties are classified as physical or chemical.

proton   A subatomic particle found in the nucleus of all atoms; has a charge of $+1$ and a mass of about 1 amu. An H^+ ion is a proton.

reactant   A chemical substance entering into a reaction.

redox   An abbreviation for *oxidation–reduction.*

reducing agent   A substance that causes a decrease in the oxidation state of another substance. The reducing agent is oxidized during the course of a reaction.

reduction   A decrease in the oxidation number of an element as a result of gaining electrons.

salts   Ionic compounds of cations and anions.

saturated solution   A solution containing dissolved solute in equilibrium with undissolved solute.

scientific method   A method of solving problems by observation; recording and evaluating data of an experiment; formulating hypotheses and theories to explain the behavior of nature; and devising additional experiments to test the hypotheses and theories to see if they are correct.

scientific notation   A number between 1 and 10 (the decimal point after the first nonzero digit) multiplied by 10 raised to a power; for example, 6.022×10^{23}.

significant figures   The number of digits that are known plus one that is uncertain are considered significant in a measured quantity.

simplest formula   *See* empirical formula.

single bond   A covalent bond in which one pair of electrons is shared between two atoms.

single-displacement reaction   A reaction of an element and a compound to produce a different element and a different compound.

solid   One of the three physical states of matter; matter in the solid state has a definite shape and a definite volume.

solubility   An amount of solute that will dissolve.

solute   The substance that is dissolved in a solvent to form a solution.

solution   A homogeneous mixture of two or more substances.

solvent The substance present to the largest extent in a solution. The solvent dissolves the solute.

specific gravity The ratio of the density of one substance to the density of another substance taken as a standard. Water is usually the standard for liquids and solids; air, for gases.

specific heat The quantity of heat required to change the temperature of 1 gram of any substance by 1°C.

spectator ion An ion in solution that does not undergo chemical change during a chemical reaction.

standard boiling point *See* normal boiling point.

standard conditions *See* STP.

stoichiometry The area of chemistry that deals with the quantitative relationships among reactants and products in a chemical reaction.

STP (standard temperature and pressure) 0°C (273 K) and 1 atm (760 torr).

subatomic particles Mainly protons, neutrons, and electrons.

sublimation The process of going directly from the solid state to the vapor state without becoming a liquid.

substance Matter that is homogeneous and has a definite, fixed composition. Substances occur in two forms—as elements and as compounds.

symbol In chemistry, an abbreviation for the name of an element.

temperature A measure of the intensity of heat or how hot or cold a system is.

theory An explanation of the general principles of certain phenomena with considerable evidence of facts to support it.

titration The process of measuring the volume of one reagent required to react with a measured weight or volume of another reagent.

torr A unit of pressure (1 torr = 1 mm Hg).

total ionic equation An equation that shows compounds in the form in which they actually exist. Strong electrolytes are written as ions in solution, whereas nonelectrolytes, weak electrolytes, precipitates, and gases are written in the un-ionized form.

transition elements The metallic elements characterized by increasing numbers of *d* and *f* electrons in an inner shell. These elements are located in Groups IB through VIIB and in Group VIII of the periodic table.

triple bond A covalent bond in which three pairs of electrons are shared between two atoms.

unsaturated solution A solution containing less solute per unit volume than its corresponding saturated solution.

vapor pressure The pressure exerted by a vapor in equilibrium with its liquid.

vapor pressure curve A graph generated by plotting the temperature of a liquid on the *x* axis and its vapor pressure on the *y* axis. Any point on the curve represents an equilibrium between the vapor and liquid.

volume The amount of space occupied by matter.

volume percent solution The volume of solute in 100 ml of solution.

water of crystallization or hydration Water molecules that are part of a crystalline structure, as in a hydrate.

weak electrolyte A substance that is ionized to a small extent in aqueous solution.

weight (mass) An extraneous property that an object possesses. The weight of an object depends on the gravitational attraction of the earth for that object. Therefore, an object's weight depends on its location in relation to the earth.

weight percent solution The grams of solute in 100 g of a solution.

yield The amount of product obtained from a chemical reaction.

MEASUREMENT TABLES
PREFIXES AND NUMERICAL VALUES FOR SI UNITS

Prefix	Symbol	Numerical value	Power of 10 equivalent
exa	E	1,000,000,000,000,000,000	10^{18}
peta	P	1,000,000,000,000,000	10^{15}
tera	T	1,000,000,000,000	10^{12}
giga	G	1,000,000,000	10^{9}
mega	M	1,000,000	10^{6}
kilo	k	1,000	10^{3}
hecto	h	100	10^{2}
deka	da	10	10^{1}
—	—	1	10^{0}
deci	d	0.1	10^{-1}
centi	c	0.01	10^{-2}
milli	m	0.001	10^{-3}
micro	μ	0.000001	10^{-6}
nano	n	0.000000001	10^{-9}
pico	p	0.000000000001	10^{-12}
femto	f	0.000000000000001	10^{-15}
atto	a	0.000000000000000001	10^{-18}

SI UNITS AND CONVERSION FACTORS

Length
SI unit: meter (m)

1 meter	= 1.0936 yards
1 centimeter	= 0.3937 inch
1 inch	= 2.54 centimeters (exactly)
1 kilometer	= 0.62137 mile
1 mile	= 5280 feet
	= 1.609 kilometers
1 angstrom	= 10^{-10} meter

Mass
SI unit: kilogram (kg)

1 kilogram	= 1000 grams
	= 2.20 pounds
1 pound	= 453.59 grams
	= 0.45359 kilogram
	= 16 ounces
1 ton	= 2000 pounds
	= 907.185 kilograms
1 ounce	= 28.3 g
1 atomic mass unit	= 1.6606×10^{-27} kilograms

Volume
SI unit: cubic meter (m^3)

1 liter	= $10^{-3} m^3$
	= 1 dm^3
	= 1.0567 quarts
1 gallon	= 4 quarts
	= 8 pints
	= 3.785 liters
1 quart	= 32 fluid ounces
	= 0.946 liter
1 fluid ounce	= 29.6 mL

Temperature
SI unit: kelvin (K)

0 K	= –273.15°C
	= –459.67°F
K	= °C + 273.15
C°	= $\dfrac{(\text{°F} - 32)}{1.8}$
°F	= 1.8(°C) + 32
°F	= 1.8(°C + 40) –40

Energy
SI unit: joule (J)

1 joule	= 1 kg m^2/s^2
	= 0.23901 calorie
1 calorie	= 4.184 joules

Pressure
SI unit: pascal (Pa)

1 pascal	= 1 kg/$m^1 s^2$
1 atmosphere	= 101.325 kilopascals
	= 760 torr
	= 760 mm Hg
	= 14.70 pounds per square inch (psi)